Hydrogeologic Framework of the Wood River Valley Aquifer System, South-Central Idaho

By James R. Bartolino and Candice B. Adkins

Prepared in cooperation with Blaine County, City of Hailey, City of Ketchum, The Nature Conservancy, City of Sun Valley, Sun Valley Water and Sewer District, Blaine Soil Conservation District, and City of Bellevue

Scientific Investigations Report 2012–5053

U.S. Department of the Interior
U.S. Geological Survey

U.S. Department of the Interior
KEN SALAZAR, Secretary

U.S. Geological Survey
Marcia K. McNutt, Director

U.S. Geological Survey, Reston, Virginia: 2012

For more information on the USGS—the Federal source for science about the Earth, its natural and living resources, natural hazards, and the environment, visit http://www.usgs.gov or call 1–888–ASK–USGS.

For an overview of USGS information products, including maps, imagery, and publications, visit http://www.usgs.gov/pubprod

To order this and other USGS information products, visit http://store.usgs.gov

Suggested citation:
Bartolino, J.R., and Adkins, C.B., 2012, Hydrogeologic framework of the Wood River Valley aquifer system, south-central Idaho: U.S. Geological Survey Scientific Investigations Report 2012–5053, 46 p.

Contents

Abstract 1

Introduction...2

   Purpose and Scope 2

   Description of Study Area 3

      Groundwater..3

      Surface Water 3

   Previous Work 5

Study Methods 6

   Definition of Aquifer..6

   Delineation of Aquifer Boundary and Extent..6

   Analyses of Well Data 6

   Geophysical Surveys..8

   Compilation of Bedrock Surface and Sediment Thickness Maps.............................10

   Compilation of Fine-Grained Sediment Map 10

   Estimation of Hydraulic Properties 10

Hydrogeologic Framework 12

   Geologic Setting..12

      Pre-Quaternary History..12

      Quaternary History..15

   Hydrogeologic Units 15

      Paleozoic Sedimentary Rocks 15

         Ordovician and Silurian Rocks 15

         Milligen Formation (Upper to Lower Devonian)..15

         Sun Valley Group..16

            Wood River Formation (Middle Pennsylvanian to Early Permian)...................16

            Dollarhide Formation (Permian) 16

      Cretaceous Intrusive Rocks 16

      Tertiary Igneous Rocks 16

         Challis Volcanic Group (Eocene) 16

         Idavada Volcanics (Miocene)...17

      Quaternary Sedimentary Deposits and Basalt 17

         Coarse-Grained Sedimentary Deposits 17

         Fine-Grained Sedimentary Deposits..17

         Basalts and Associated Deposits 20

         Consolidated Rock Surface Underlying Quaternary Sediment and Thickness of Quaternary Sediment 21

         Hydraulic Conductivity..23

   Groundwater Movement 23

      Hydrogeologic Controls on Groundwater/Surface-Water Interaction............26

      Subsurface Groundwater Flow Beneath Silver Creek and Stanton Crossing................26

      Bedrock Flow Systems and Thermal Springs 27

Contents—Continued

Data Needs and Suggestions for Further Study 27

Summary and Conclusions 28

Acknowledgments 30

References Cited 30

Glossary 37

Appendix A. Well Information Used in the Estimation of the Altitude of the Pre-Quaternary
Bedrock Surface and Top of Quaternary Basalt, Wood River Valley Aquifer System,
South-Central Idaho 45

Appendix B. Horizontal-to-Vertical Spectral Ratio Measurements Used to Estimate
Altitude of the Pre-Quaternary Bedrock Surface and Top of the Quaternary Basalt,
Wood River Valley Aquifer System, South-Central Idaho 45

Appendix C. Well Information For Wells Not Completed to the Pre-Quaternary Bedrock
Surface or Top of Quaternary Basalt, Wood River Valley Aquifer System,
South-Central Idaho 45

Appendix D. Well Information Used in the Estimation of the Top and Corresponding
Thickness of the Uppermost Unit of Fine-Grained Sediment Within the Wood River
Valley Aquifer System, Southern Wood River Valley, South-Central Idaho 45

Appendix E. Estimates of Transmissivity and Hydraulic Conductivity for 81 Wells in the
Wood River Valley, South-Central Idaho 45

Plate

Plate 1. Estimated altitude of pre-Quaternary bedrock surface and top of Quaternary basalt, Wood River Valley aquifer system, south-central Idaho

Figures

Figure 1. Map showing locations of communities and selected U.S. Geological Survey stream-gaging stations, wells, a Horizontal-to-Vertical Spectral Ratio measurement point, hot springs, and features, Wood River Valley, south-central Idaho 4

Figure 2. Diagram showing well-numbering system used in Idaho 7

Figure 3. Graphs showing representative Horizontal-to-Vertical Spectral Ratio (HVSR) data collected in the Wood River Valley, south-central Idaho 9

Figure 4. Generalized geologic map of the Wood River Valley and surrounding area, south-central Idaho 14

Figure 5. Hydrogeologic sections showing generalized lithologic units 18

Figure 6. Map showing estimated altitude of the top and corresponding thickness of the uppermost unit of fine-grained sediment within the Wood River Valley aquifer system, southern Wood River Valley, south-central Idaho 19

Figure 7. Map showing estimated thickness of Quaternary sediment in the Wood River Valley aquifer system, Wood River Valley, south-central Idaho 22

Figure 8. Map showing estimated hydraulic conductivity for selected wells completed in the Wood River Valley aquifer system and pre-Quaternary bedrock, Wood River Valley, south-central Idaho 24

Tables

Table 1. Geologic time scale with hydrogeologic units of the Wood River Valley area and significant orogenies and geologic events 13

Table 2. Summary statistics, estimated hydraulic conductivity of 81 wells in the Wood River Valley aquifer system by the Thomasson and others (1960) and Theis and others (1963) methods 25

Table 3. Published ranges of hydraulic conductivity for unconsolidated sediment and basalt 25

Conversion Factors, Datums, and Abbreviations and Acronyms

Conversion Factors

Multiply	By	To obtain
foot (ft)	0.3048	meter (m)
mile (mi)	1.609	kilometer (km)
square foot (ft^2)	0.09290	square meter (m^2)
square mile (mi^2)	2.590	square kilometer (km^2)
cubic foot (ft^3)	0.02832	cubic meter (m^3)
acre-foot (acre-ft)	1,233	cubic meter (m^3)
acre-foot per year (acre-ft/y)	1,233	cubic meter per year (m^3/y)
foot per day (ft/d)	0.3048	meter per day (m/d)
foot squared per day (ft^2/d)	0.09290	meter squared per day (m^2/d)

*Transmissivity: The standard unit for transmissivity is cubic foot per day per square foot times foot of aquifer thickness [(ft^3/d)/ft^2]ft. In this report, the mathematically reduced form, foot squared per day (ft^2/d), is used for convenience.

Datums

Vertical coordinate information is referenced to the North American Vertical Datum of 1988 (NAVD 88).

Horizontal coordinate information is referenced to the North American Datum of 1983 (NAD 83).

Altitude, as used in this report, refers to distance above the vertical datum.

Abbreviations and Acronyms

Abbreviation or acronym	Definition
GIS	Geographic Information System
HVSR	Horizontal-to-Vertical Spectral Ratio
IDWR	Idaho Department of Water Resources
PLSS	Public Land Survey System
USGS	U.S. Geological Survey

Hydrogeologic Framework of the Wood River Valley Aquifer System, South-Central Idaho

By James R. Bartolino and Candice B. Adkins

Abstract

The Wood River Valley contains most of the population of Blaine County and the cities of Sun Valley, Ketchum, Hailey, and Bellevue. This mountain valley is underlain by the alluvial Wood River Valley aquifer system, which consists primarily of a single unconfined aquifer that underlies the entire valley, an underlying confined aquifer that is present only in the southernmost valley, and the confining unit that separates them. The entire population of the area depends on groundwater for domestic supply, either from domestic or municipal-supply wells, and rapid population growth since the 1970s has caused concern about the long-term sustainability of the groundwater resource. As part of an ongoing U.S. Geological Survey effort to characterize the groundwater resources of the Wood River Valley, this report describes the hydrogeologic framework of the Wood River Valley aquifer system.

Although most of the Wood River Valley aquifer system is composed of Quaternary-age sediments and basalts of the Wood River Valley and its tributaries, older igneous, sedimentary, or metamorphic rocks that underlie these Quaternary deposits also are used for water supply. It is unclear to what extent these rocks are hydraulically connected to the main part of Wood River Valley aquifer system and thus whether they constitute separate aquifers. Paleozoic sedimentary rocks in and near the study area that produce water to wells and springs are the Phi Kappa and Trail Creek Formations (Ordovician and Silurian), the Milligen Formation (Devonian), and the Sun Valley Group including the Wood River Formation (Pennsylvanian-Permian) and the Dollarhide Formation (Permian). These sedimentary rocks are intruded by granitic rocks of the Late Cretaceous Idaho batholith. Eocene Challis Volcanic Group rocks overlie all of the older rocks (except where removed by erosion). Miocene Idavada Volcanics are found in the southern part of the study area. Most of these rocks have been folded, faulted, and metamorphosed to some degree, thus rock types and their relationships vary over distance.

Quaternary-age sediment and basalt compose the primary source of groundwater in the Wood River Valley aquifer system. These Quaternary deposits can be divided into three units: a coarse-grained sand and gravel unit, a fine-grained silt and clay unit, and a single basalt unit. The fine- and coarse-grained units were primarily deposited as alluvium derived from glaciation in the surrounding mountains and upper reaches of tributary canyons. The basalt unit is found in the southeastern Bellevue fan area and is composed of two flows of different ages. Most of the groundwater produced from the Wood River Valley aquifer system is from the coarse-grained deposits.

The altitude of the pre-Quaternary bedrock surface in the Wood River Valley was compiled from about 1,000 well-driller reports for boreholes drilled to bedrock and about 70 Horizontal-to-Vertical Spectral Ratio (HVSR) ambient-noise measurements. The bedrock surface generally mimics the land surface by decreasing down tributary canyons and the main valley from north to south; it ranges from more than 6,700 feet in Baker Creek to less than 4,600 feet in the central Bellevue fan. Most of the south-central portion of the Bellevue fan is underlain by an apparent topographically closed area on the bedrock surface that appears to drain to the southwest towards Stanton Crossing. Quaternary sediment thickness ranges from less than a foot on main and tributary valley margins to about 350 feet in the central Bellevue fan.

Hydraulic conductivity for 81 wells in the study area was estimated from well-performance tests reported on well-driller reports. Estimated hydraulic conductivity for 79 wells completed in alluvium ranges from 1,900 feet per day (ft/d) along Warm Springs Creek to less than 1 ft/d in upper Croy Canyon. A well completed in bedrock had an estimated hydraulic conductivity value of 10 ft/d, one well completed in basalt had a value of 50 ft/d, and three wells completed in the confined system had values ranging from 32 to 52 ft/d.

Subsurface outflow of groundwater from the Wood River Valley aquifer system into the eastern Snake River Plain aquifer was estimated to be 4,000 acre-feet per year. Groundwater outflow beneath Stanton Crossing to the Camas Prairie was estimated to be 300 acre-feet per year.

Introduction

The population of Blaine County in south-central Idaho has nearly quadrupled—from about 5,700 to 22,000 people—from 1970 to 2010 (Forstall, 1995; U.S. Census Bureau, 2011). In addition to permanent residents, thousands of people annually visit Blaine County for winter and summer recreation. Most population growth and recreational use is in the northernmost part of the county in the Wood River Valley. The entire population of the valley depends on groundwater for domestic supply, either from privately owned or municipal-supply wells; surface water is used for recreation and irrigation.

Water managers and private landowners are increasingly concerned about the effects of population growth on groundwater and surface-water supplies in the area, particularly the sustainability of groundwater resources and the effects of wastewater disposal on the quality of both sources. Development in recent years has moved into tributary canyons of the Wood River Valley, and residents in some of these canyon areas have reported declining groundwater levels. It is uncertain whether the declining water levels are caused by pumping that has accompanied increased development or are a response to several years of drought conditions. [In Idaho's climate division 4, 61 percent of the months between October 1995 and September 2011 were classified as being under drought conditions as defined by a Palmer Drought Severity Index of less than -0.5 (National Climatic Data Center, 2011)]. In June 2005, Blaine County Commissioners approved an interim moratorium on selected development activities while the effects of growth, including those on water resources, were evaluated.

Although several studies and the resulting technical reports have addressed specific water-related issues or aspects in selected areas of the Wood River Valley, a current, comprehensive evaluation of water resources in the valley is needed to address concerns about the effects of current development and the potential effects of continued growth and development. In 2005, the U.S. Geological Survey (USGS), in cooperation with several local government agencies and organizations, completed a compilation and review of existing information and data on the hydrology of the Wood River Valley, identified gaps in information about water resources, and proposed a work plan with priorities for data collection and interpretation to fill these gaps. The objectives of the overall work plan for the USGS study are: (1) to provide data and interpretations about the water resources of the Wood River Valley to enable county and local governments to make informed decisions about the management of those resources, (2) to identify any additional water-resources data collection or analyses that would assist decision makers, and (3) to construct a hydrogeologic framework for the Wood River Valley. The first phase of the work plan, compilation of groundwater-level maps for partial-development and 2006 conditions, the change between them, and an analysis of hydrologic trends in the groundwater and surface-water systems was completed in 2007 (Skinner and others, 2007). (Ideally, water levels measured prior to 1970 would be used to construct a predevelopment map, however, because there are insufficient pre-1970 water-level measurements to adequately represent the entire Wood River Valley aquifer system, water-level data collected as recently as 1986 was selectively used to expand map coverage into recently developed areas.) The second phase of the plan, development of a groundwater budget, was completed in 2009 (Bartolino, 2009). The third phase of the 2005 work plan, construction of a hydrogeologic framework, is described in this report.

Purpose and Scope

This report describes the development of an updated hydrogeologic framework for the Wood River Valley aquifer system, including some areas that have not been addressed by previous investigators and authors. The updated framework is based on a review and interpretation of 3,000 drilling reports, geologic maps, previous work, and geophysical surveys conducted for the current work. The description of the hydrogeologic framework includes: (1) the approximate altitude of the bedrock surface that marks the lower boundary of the aquifer system and the top of Quaternary-age basalts that are part of the aquifer system, (2) the occurrence of Quaternary-age coarse- and fine-grained alluvium and basalt that constitute the Wood River Valley aquifer system, and (3) the distribution of hydraulic conductivity within the aquifer system. This report also includes brief descriptions of the bedrock units underlying the aquifer system that locally provide or potentially provide water to wells located primarily in tributary canyons and along valley margins. The hydrogeology of these older bedrock units is described in some detail because an increasing number of wells in the Wood River Valley are completed in these older units, which may have both different water-quality characteristics and sources of water than the Wood River Valley aquifer system. A glossary is included with definitions for selected technical terms contained in the report.

Description of Study Area

The Wood River Valley of south-central Idaho extends from Galena Summit (at the head of the Big Wood River drainage north of the study area) southward to the Picabo and Timmerman Hills (fig. 1). The valley can be separated into upper and lower parts along an east-west line immediately south of Bellevue: the upper valley is narrow, broadening downstream to a maximum width of 2 mi and the lower valley opens into a triangular alluvial fan (the Bellevue fan) about 9 mi across at its southern end. The study area of this report is the part of the Wood River Valley aquifer system that extends from Baker Creek, southward to the Timmerman Hills and is of greater areal extent (100 mi^2) than that defined by Skinner and others (2007) (86 mi^2) and used by Bartolino (2009).

The Wood River Valley has a relatively flat bottom, and land-surface altitudes range from about 6,800 ft at the northern boundary of the study area near the confluence of the river with Baker Creek to about 4,800 ft at the southern boundary near Picabo. A number of tributary canyons intersect the valley, the largest of which are those of North Fork Big Wood River, Warm Springs Creek, Trail Creek, East Fork Big Wood River, Deer Creek, and Croy Creek (fig. 1). The main valley and the tributary canyons have steep sides and are surrounded by highlands with peaks that reach altitudes of more than 11,000 ft.

In addition to their different physiographic characteristics, the upper and lower valleys also differ in land use. The upper Wood River Valley is more developed and contains the incorporated communities of Sun Valley, Ketchum, Hailey, and Bellevue. Land use in the upper valley is predominantly residential with many large homes situated on landscaped acreage. The lower Wood River Valley is dominated by farms and ranches (irrigated by groundwater and diverted surface water), and contains the small communities of Gannett and Picabo. Although some of the tributary canyons in the upper valley, such as Trail and Warm Springs Creeks, have supported development for more than 50 years, more recent development has expanded into the valley's other tributary canyons. Three wastewater-treatment plants in the study area discharge to the Big Wood River, but many homes rely on septic systems for wastewater disposal. A more complete description of the study area, including climate, is available in Skinner and others (2007).

Groundwater

The Wood River Valley aquifer system is composed of a single unconfined aquifer that underlies the entire valley, an underlying confined aquifer that is present only to the south of Baseline Road (fig. 1), and the confining unit separating the two aquifers. The aquifer system consists primarily of Quaternary sediments of the Wood River Valley, although Quaternary basalts and interbedded sediment are the primary source of water in the southeastern Bellevue fan.

The confined aquifer is separated from the overlying unconfined aquifer by fine-grained lacustrine deposits. This confining unit thickens towards the south and, generally, as land-surface altitude decreases in the same direction, the potentiometric surface rises above land surface and some wells flow under artesian pressure. South of Gannett, near the westernmost extent of Quaternary basalt, the confining unit fingers out and the aquifer becomes solely unconfined again.

Depth to groundwater in the upper valley is commonly less than 10 ft, and increases to approximately 90 ft southward. Water levels in wells completed in the unconfined aquifer in the lower valley range from less than 10 ft to approximately 150 ft, whereas wells completed in the confined aquifer are under artesian pressure and flow where the potentiometric surface is above land surface (Skinner and others, 2007).

Surface Water

Most of the Wood River Valley is drained by the Big Wood River or its tributaries, except for the southeastern portion of the Bellevue fan, which is drained by Silver Creek, a tributary to the Little Wood River. The Big Wood and Little Wood Rivers meet near Gooding, approximately 35 mi southwest of the study area, where they become the Malad River, a tributary to the Snake River. The Big Wood River originates near Galena Summit, approximately 20 mi northwest of Ketchum, and it gains flow from a number of perennial and ephemeral tributaries as it meanders across the narrow upper valley. At Bellevue, the channel follows the western side of the Bellevue fan (although flow through most of this reach is ephemeral), finally exiting the valley at the Big Wood River at Stanton Crossing near Bellevue gaging station (13140800) (fig. 1). Fed by springs and seeps, Silver Creek and its tributaries originate on the Bellevue fan and flow out of the valley at Picabo (fig. 1). Most of the streams in the tributary canyons to the Big Wood River are ephemeral and flow only in response to precipitation or snowmelt; however, North Fork Big Wood River, Warm Springs Creek, Trail Creek, East Fork Big Wood River, Deer Creek, and Croy Creek typically flow into the Big Wood River year-round. Streams in some of the smaller tributary canyons are perennial in their upper reaches, and some of this water likely infiltrates directly into the aquifer system or reaches the Big Wood River by subsurface flow through streambed gravels. Most of the Wood River Valley was under irrigation by 1900 (Jones, 1952), and a well-developed network of irrigation canals and drains exists throughout the study area. The diversions and return flows between the irrigation system and the Big Wood River, as well as the exchange of water between the canals, drains, and streams and the underlying unconfined aquifer complicate the interpretation of streamflow measurements.

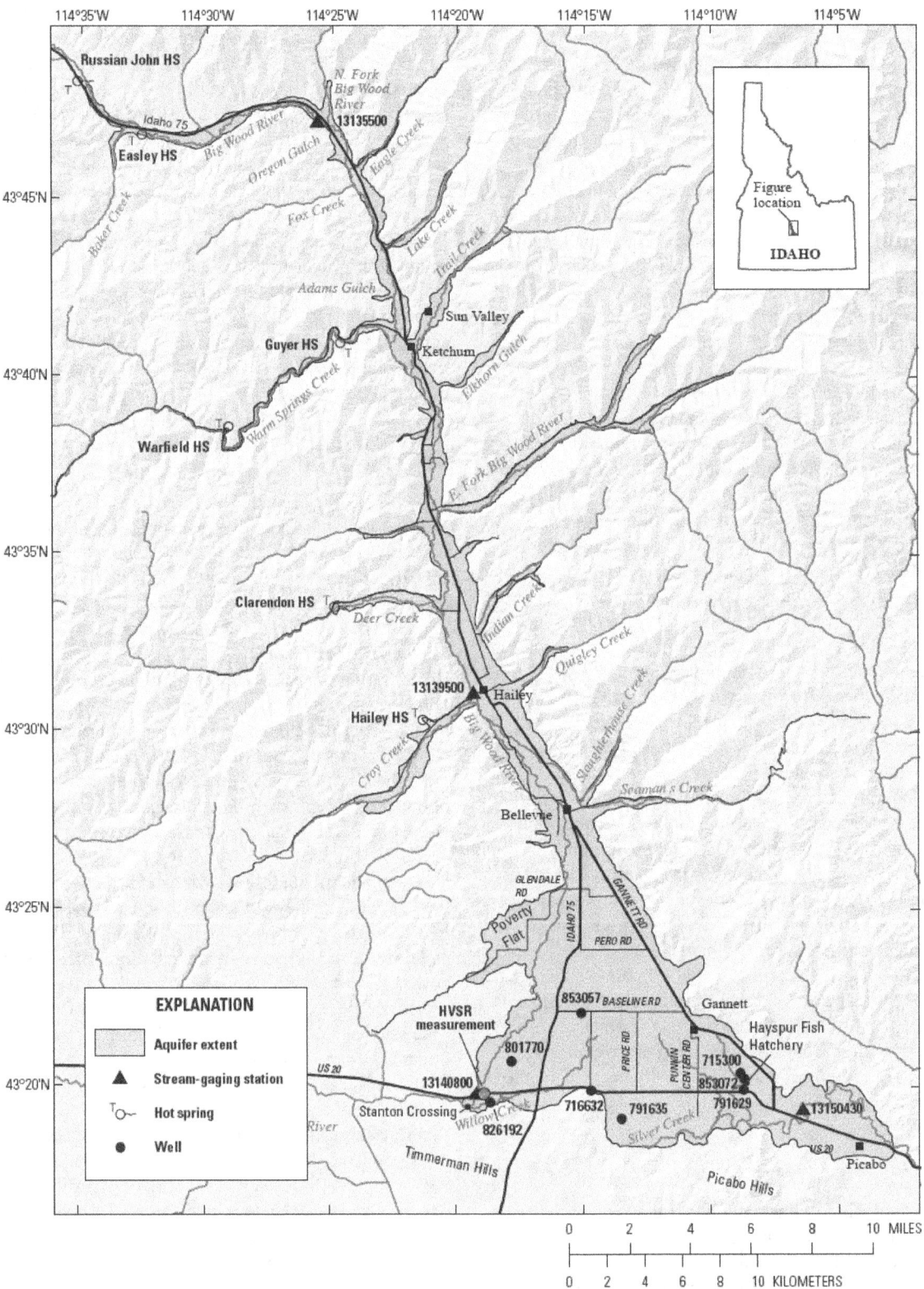

Figure 1.   Locations of communities and selected U.S. Geological Survey stream-gaging stations, wells, a Horizontal-to-Vertical Spectral Ratio measurement point, hot springs, and features, Wood River Valley, south-central Idaho.

Previous Work

The earliest publications on the geology of the study area focus on the mining of precious metals discovered in the area in 1864 (Spence, 1999). These early reports include those of Eldridge (1896), Lindgren (1900), and Umpleby and others (1930). Although the geology of the study area is described in these reports, no mention is made of water resources.

The earliest papers that discuss groundwater in the Wood River Valley are those of Chapman (1921) and Stearns and others (1938); however, Smith (1959) made the first detailed study of the Wood River Valley aquifer system in the Hailey and the Bellevue fan areas. In characterizing the aquifer system, Smith (1959) made multiple-well aquifer tests at five sites and constructed geologic sections of the aquifer in the area of four USGS streamflow-gaging stations then in operation in the lower valley. Smith (1960) also analyzed groundwater underflow at 27 streamflow-gaging stations in the Big Wood River basin above its confluence with the Malad River. For each streamflow-gaging station included in this latter report, Smith (1960) described the geology (including a geologic section) and made an estimate of probable groundwater underflow beneath the stream channel.

Schmidt (1962) described the Quaternary geologic history and processes that formed the Wood River Valley south of Hailey and the eastern part of the Camas Prairie. This work has served as the basis for all subsequent descriptions of the Quaternary rocks of the study area although later authors have expanded and refined Schmidt's original description.

Castelin and Chapman (1972) summarized and updated previous reports on water resources of the Wood River Valley and collected a limited amount of new data. Castelin and Winner (1975) addressed the effects of urbanization on water resources in the Sun Valley-Ketchum area. As part of their aquifer characterization of the Wood River Valley, Castelin and Winner constructed four geologic sections of the area between Hailey and the North Fork of the Big Wood River on the basis of well-driller reports and a single seismic refraction survey at their northernmost section across the Big Wood River above the confluence with the North Fork.

Moreland (1977) focused on groundwater and surface-water relations in the Silver Creek area of the southeastern Bellevue fan. The report contains four geohydrologic sections, descriptions of the lithology of 10 test holes drilled for the study, maps of the extent of the confining unit separating the confined and unconfined aquifers, and estimates of transmissivity (using single-well, specific capacity methods) from about 70 well-driller reports.

Luttrell and Brockway (1984) studied the effects of septic systems on water quality in the Wood River Valley between the North Fork of the Big Wood River and 3 mi south of Bellevue. They included two generalized geologic sections, one for the northern part of their study area and one for the southern part.

Frenzel's (1989) report on the water resources of the Wood River Valley contains a geologic section of the aquifer system near Glendale Road constructed from seismic refraction data. This report also includes estimates of hydraulic conductivity (using single-well, specific capacity methods) at six wells using pumping data from well reports and Smith's (1959) multiple-well aquifer tests.

Brockway and Kahlown (1994) compiled hydrogeologic information from previous authors in preparation for constructing a groundwater-flow model of the Wood River Valley aquifer system south of Hailey; the model is described in Wetzstein and others (1999). The Wetzstein and others (1999) report updated the hydrogeologic framework described in Brockway and Kahlown (1994) by incorporating additional data from well-driller reports. Brown (2000) "synthesize[ed] and highlight[ed] the methods and findings" of the reports by Brockway and Kahlown (1994) and Wetzstein and others (1999).

Various geologic maps of the study area have been compiled, including regional-scale geologic maps (Worl and others, 1991; Link and Rodgers, 1995), a geologic terrane map (Worl and Johnson, 1995b), and a sedimentary unit map (Link and others, 1995b). The Link and Rodgers (1995) map includes three northeast-trending geologic sections that cross the Wood River Valley: (1) 2 mi north of Ketchum, (2) 1 mi north of Hailey, and (3) 1 mi north of Bellevue. These sections emphasize the regional structure and in places extend to depths of as much as 2 mi below land surface. Recent mapping products of the area by the Idaho Geological Survey include a surficial geologic map (Breckenridge and Othberg, 2006) and 1:24,000-scale geologic maps that include the Magic Reservoir East quadrangle (Kauffman and Othberg, 2007), the Gannett quadrangle (Garwood and others, 2010), and the Picabo quadrangle (Garwood and others, 2011). The map by Garwood and others (2010) includes the subsurface extent of the Pleistocene basalts and a geologic section across the Bellevue fan. Currently (2012), the Idaho Geological Survey has plans to map the Bellevue and Seaman's Creek 1:24,000-scale quadrangles for ultimate inclusion into a geologic map of the Fairfield 1:100,000-scale quadrangle.

A description of previous work on the hydrology of the Wood River Valley area related to groundwater-level maps and surface-water flows is available in Skinner and others (2007). Similarly, previously published studies that developed water budgets for all or part of the Wood River Valley are described in Bartolino (2009).

Study Methods

The hydrogeologic framework described in the current report incorporates several different components. These components are based on the interpretation and analysis of the work of previous authors, analysis of well-driller reports, and new data including geophysical surveys. Much of the analysis and the construction of all of the maps involved the use of a geographic information system (GIS).

Definition of Aquifer

This report uses "Wood River Valley aquifer system" as defined by Skinner and others (2007, p. 4):

> "…the aquifer system of the Wood River Valley is here informally named the Wood River Valley aquifer system. It includes the Quaternary sediments of the Wood River Valley and its tributaries and locally, underlying igneous, sedimentary, or metamorphic rocks where they are hydraulically connected and used for water supply. The Wood River Valley aquifer system consists of a single unconfined aquifer and an underlying confined aquifer present south of Baseline Road."

This definition implicitly includes the Quaternary-age basalts of the Gannett and Picabo area. In the present report, however, these basalts are explicitly included as part of the aquifer system and are discussed in section, "Basalts and Associated Deposits."

Currently (2012), little is known about the water-bearing characteristics and hydraulic connection of the bedrock units underlying the Quaternary sediments and basalts of the confined and unconfined aquifers. Where they are hydraulically connected to the sediments and basalts, however, they likely represent a very small percentage of available water in the aquifer system, and Skinner and others (2007) and Bartolino (2009) generally neglected these bedrock units in their discussions.

Delineation of Aquifer Boundary and Extent

The boundary of the Wood River Valley aquifer system used in this report has been expanded from that defined by Skinner and others (2007): It has been extended farther up several tributary canyons, extended up the Big Wood River to include a number of wells near the mouth of Baker Creek, expanded to include the western side of the Poverty Flat area, and extended slightly downstream of Stanton Crossing and Picabo. Although most of the boundary delineated by Skinner and others (2007) has been retained here, modifications were made in places to include information from additional wells that penetrated bedrock below alluvium.

Skinner and others (2007, p. 14) described the methodology used to define their boundary thus:

> "The study area boundary was defined by using a [digital-elevation model] converted to a slope model to determine the transition from the adjacent bedrock hills (high slope values) to the flat portions of the valley filled with unconsolidated sediment (low slope values). This transition represents the approximate location of the boundary between the unconfined aquifer and the adjacent bedrock in the tributary canyons and the main valley."

For the current work, the aquifer boundaries were extended visually along the transition from steeper hillsides to relatively flat valley bottoms. As with the Skinner and others (2007) boundary, the extended boundary corresponds to the contact between Quaternary alluvial deposits and surrounding consolidated rocks mapped by Worl and others (1991) and Breckenridge and Othberg (2006).

Analyses of Well Data

The primary sources of data for this hydrogeologic framework of the Wood River Valley aquifer system are the well-driller reports maintained by the Idaho Department of Water Resources (IDWR) and available through an online database of Well Driller Reports (Idaho Department of Water Resources, 2012). The IDWR database contains "most of the well-driller reports dating back to July 1987" but because such reports were requested but not required by IDWR prior to 1953 (Castellin and Winner, 1975), the database does not contain reports for all wells drilled in the study area. The database does include, however, reports for many wells drilled before 1987 (the oldest well found in the database for the study area was drilled in 1950).

For the current study, about 2,900 well-driller reports (also known as drillers' logs) were retrieved from the IDWR database for the study area and immediate vicinity. Well locations on the reports have historically been reported using the Public Land Survey System (PLSS) to the 160-acre, 40-acre, or 10-acre tract level. For this study, in the absence of a more precise location, a latitude and longitude was assigned to the center of the smallest assigned tract to denote the well location. Consequently, reported well locations may vary from actual locations. The quality of location and lithologic information from well-driller reports can be highly variable.

Individual wells referred to in the text are indicated both by the IDWR permit number and a PLSS-based well location used by IDWR and the USGS, thus removing ambiguity for multiple wells at a single location. For example, 54N 04W 31DDD1 (fig. 2) designates a well in Township 54 North and Range 4 West, north and west of the Boise Base Line and Meridian, respectively; this 36 square mile area is the township. The numbers following the township indicate the section (31) within the township; the letters following the section indicate the quarter section (160-acre tract), quarter-quarter section (40-acre tract), and quarter-quarter-quarter section (10-acre tract). Quarter sections are designated by the letters A, B, C, and D in counterclockwise order from the northeast quarter of each section. Within the quarter sections, 40-acre and 10-acre tracts are lettered in the same manner. For example, well 54N 04W 31DDD1 is in the SE ¼ of the SE ¼ of the SE ¼ of section 31. If multiple wells are present within the same tract, a number following the letters (1) designates the sequential number of the well.

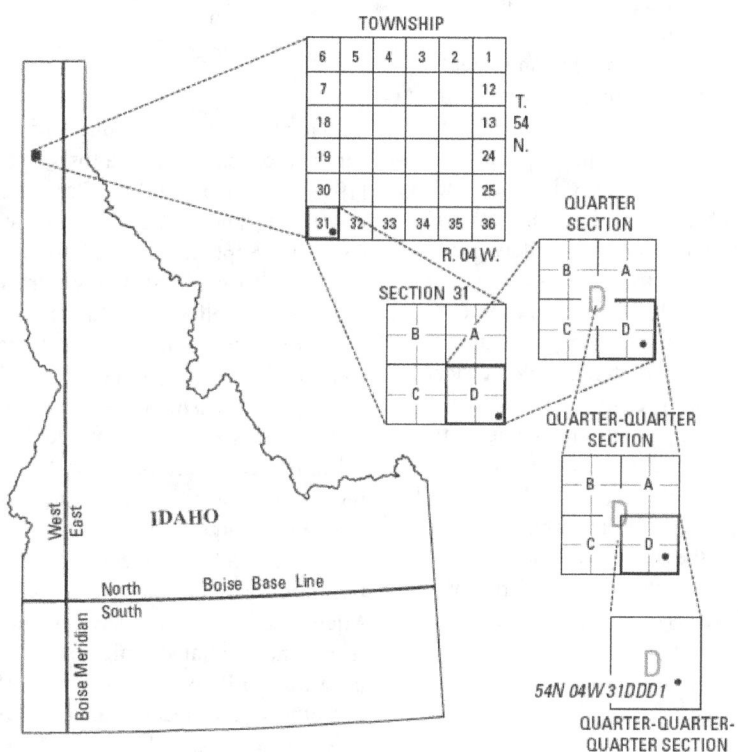

Figure 2. Well-numbering system used in Idaho.

Geophysical Surveys

Because of high productivity and increased saturated thickness of the aquifer system in the area south of Hailey, few wells penetrate to the base of the system where it is in contact with the underlying bedrock. As described in section, "Previous Work," earlier studies have made seismic refraction surveys in order to estimate the depth to the bedrock contact. For the current work, a relatively new technique was used to determine bedrock depth at a point: the Horizontal-to-Vertical Spectral Ratio (HVSR) ambient-noise method, also known as passive seismic, microtremor measurement, or seismic noise. Unlike other seismic techniques that require an artificial energy source such as explosives, weight drops, or hammer blows, the HVSR method uses a three-component seismometer to record ambient seismic noise. Such noise may be from wind, moving water, microtremors, or even certain manmade sources. The HVSR method was proposed by Nakamura (1989) and gained wide acceptance in Japan and Europe by 2002 (Parolai and others, 2002; Lane and others, 2008). Until recently, however, it has not been widely used in the United States.

As previously mentioned, the HVSR method uses a three-component broadband seismometer to measure the seismic amplitude in the north-south, east-west, and vertical directions. The amplitude spectrum is calculated for each component, the two horizontal components are averaged, the ratio of the horizontal to vertical spectrum is calculated, and a seismic resonance frequency is identified (fig. 3). For a simple two-layer earth model in which the lower layer has a seismic impedance at least twice that of the upper layer (for example, unconsolidated sediment over bedrock) the seismic resonance frequency is related to the thickness of the upper layer. This relation is expressed by equation 1:

$$f_{r0} = V_s / 4Z, \qquad (1)$$

where
   f_{r0} is fundamental seismic resonance frequency in hertz (Hz),
   V_s is average shear-wave velocity of the upper layer in meters per second (m/s), and
   Z is sediment thickness in meters (m).

Sediment thickness is then related to the fundamental seismic resonance frequency through the nonlinear regression of data from measurement sites where sediment thickness is known (equation 2):

$$Z = af_{r0}{}^{b}, \qquad (2)$$

where
   Z is sediment thickness in meters (m),
   f_{r0} is the fundamental seismic resonance frequency in hertz (Hz),
   a is a regression fitting parameter in meters (m), and
   b is a regression fitting parameter, dimensionless.

Ideally, regression equations are developed for each survey area; examples are Ibs-von Seht and Wohlenberg (1999) ($a=96$, $b=-1.388$), Parolai and others (2002) ($a=108$, $b=-1.551$), and J.W. Lane (U.S. Geological Survey, written commun., September 15, 2009) ($a=94$, $b=-1.324$).

About 130 HVSR measurements were made in the Wood River Valley, including quality assurance and repeat measurements at the same site. Four HVSR instruments were used: Three separate Guralp EDU-T low frequency 3-component seismometers connected to a laptop computer and one freestanding Micromed S.p.A Tromino ENGY seismometer (Micromed uses the term "digital tromograph" instead of seismometer). The data were analyzed using the free open-source Geopsy software suite (Di Giulio and others, 2006; Wathelet and others, 2008; Wathelet, 2011) and the Grilla software package (Castellero and Mulargia, 2009; Micromed S.p.A., 2011). Sediment thickness was interpreted from fundamental seismic resonance using the regression equations of Ibs-von Seht and Wohlenberg (1999), Parolai and others (2002), and Lane and others (2008); the depth was taken as the mean value of the three equations. A regression equation developed for the Wood River Valley was not used because too few readings were made close to wells penetrating to bedrock and the resulting equation was not considered statistically robust. Results of the HVSR surveys are discussed in section, "Consolidated Rock Surface Underlying Quaternary Sediment and Thickness of Quaternary Sediment."

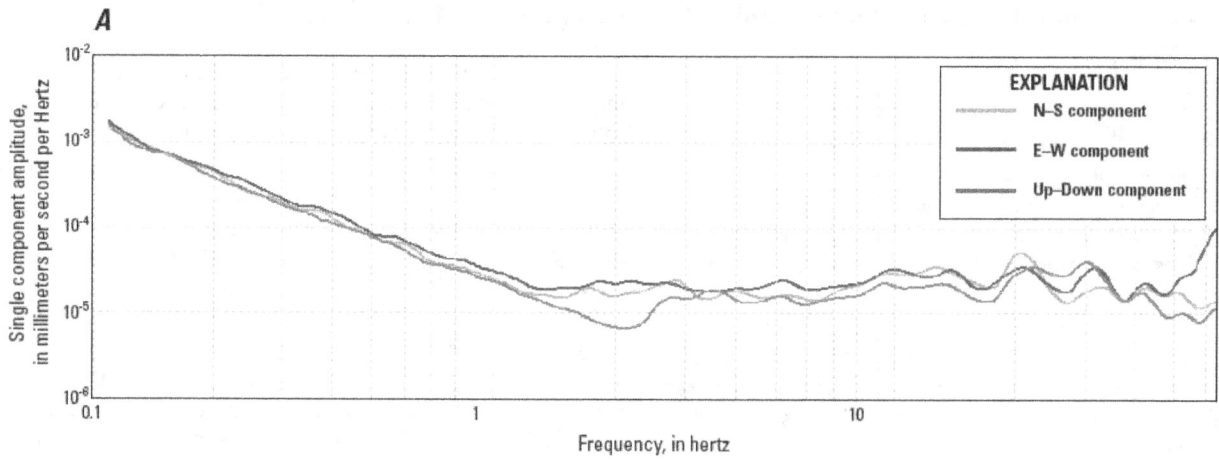

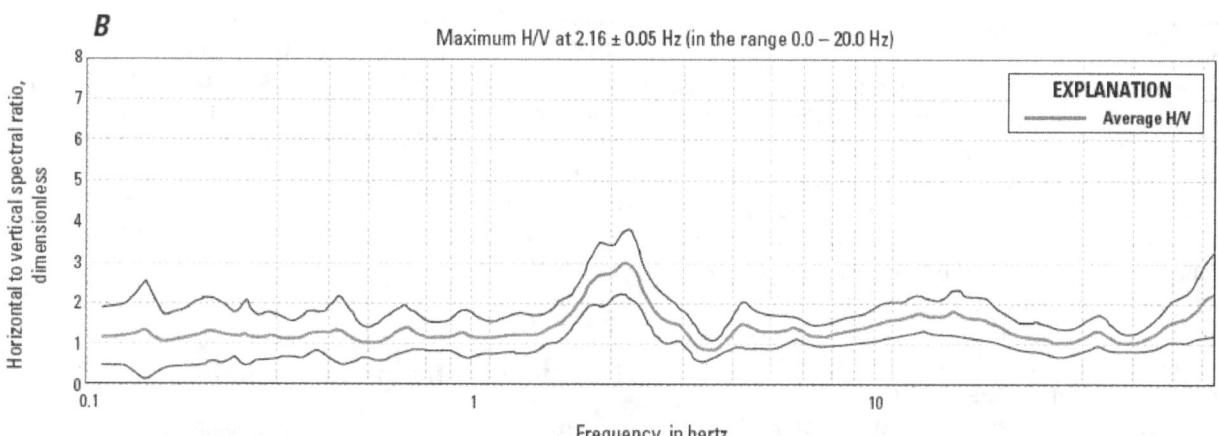

Figure 3.   Representative Horizontal-to-Vertical Spectral Ratio (HVSR) data collected in the Wood River Valley, south-central Idaho. (*A*) 3-component ambient noise seismic record. Note the separation between the horizontal and vertical amplitude around 2 hertz. (*B*) Corresponding horizontal/vertical ratio spectral plot. The main peak seismic resonance frequency is about 2.2 hertz. Graphs were generated using the Grilla software package (Castellero and Mulargia, 2009; Micromed S.p.A., 2011).

Compilation of Bedrock Surface and Sediment Thickness Maps

A contour map of the estimated altitude of the pre-Quaternary bedrock surface was compiled from about 1,000 well-driller reports for boreholes drilled to bedrock (appendix A) and about 70 HVSR measurements (appendix B); about an additional 1,500 well-driller reports for boreholes penetrating only alluvium were used to confirm and constrain contours in areas lacking bedrock wells (appendix C). About 400 well-driller reports were excluded from the analysis because of anomalous location or lithologic information or because their locations were not within or near the aquifer boundary. The altitude of the bedrock surface was estimated by subtracting the bedrock depth (taken from the well-driller reports) from the land-surface altitude (obtained from a 10-m digital-elevation-model).

Identification of the alluvium/bedrock contact is commonly subjective for several reasons: Bedrock lithology varies within and between geologic units; the bedrock surface may be heavily weathered, and lithologic descriptions differ by driller or drilling method. These factors, in combination with imprecise well location, introduce uncertainty in the determination of the actual paleotopography of the buried bedrock surface. In areas with multiple, closely spaced wells, such as the lower valley of East Fork Big Wood River, the comparison of bedrock altitude between individual wells may yield inconsistencies of up to 10 to 20 ft; however, when viewed in the context of the aquifer system as a whole, the calculated bedrock altitudes are consistent.

Several deterministic and geostatistical techniques [kriging, inverse distance weighted, natural neighbor, and spline (ESRI, 2001)] were used to interpolate the bedrock altitude surface from the well-driller reports and HVSR measurements. However, the surfaces generated using these techniques were judged either too generalized (too many well values did not correspond to the value of the interpolated surface) or overly precise (although most well values corresponded to the value of the interpolated surface, the surface configuration was deemed more complex than geologically possible). These issues probably were the result of bedrock altitude uncertainty (as discussed in the previous paragraph), the large number and high density of wells, and the irregular shape of the main valley and tributary canyons. Consequently, the interpolated surfaces were used as a guide to manually contour the altitude of the bedrock surface while honoring all points. A raster surface was then interpolated from these contours. The estimated altitude of the bedrock surface map is shown on plate 1 and discussed in section, "Consolidated Rock Surface Underlying Quaternary Sediment and Thickness of Quaternary Sediment."

Compilation of Fine-Grained Sediment Map

A map of the estimated altitude of the top of the fine-grained sediment and corresponding thickness in the southern Bellevue fan was compiled from data in about 160 well-driller reports for boreholes drilled within the area of the approximate extent of the confined aquifer (appendix D). As described in the previous section, "Compilation of Bedrock Surface and Sediment Thickness Maps," descriptions of material penetrated in a borehole may widely vary between drillers. In addition, mud- and air-rotary drilling methods tend to cause an underestimation of the amount of fine-grained sediment, especially in saturated sediments. These effects, in combination with natural variability, cause estimates of the depth and thickness of the fine-grained sediment that makes up the confining unit to be extremely variable. In order to map these sediments, the altitude of the shallowest fine-grained sediment in each well was used to generate a surface by kriging methods. (The surface was generated by using ordinary kriging, a spherical semivariogram model, a variable search radius, and the 80 nearest points, thus producing a smoothed surface.) Because of the extreme variability in the recorded thickness of the fine-grained sediments on well-driller reports, the thickness of the uppermost fine-grained unit was indicated at each point representing a well. This map is discussed in section, "Fine-Grained Sedimentary Deposits."

Estimation of Hydraulic Properties

Because few multiple-well aquifer tests have been made in the Wood River Valley aquifer system, the current report uses well-performance (pump-test or specific-capacity test) data from well-driller reports to estimate transmissivity and hence hydraulic conductivity. A number of techniques have been developed to make such estimates; the reader is referred to Mace (2001) for a comprehensive survey of analytical, empirical, and geostatistical methods and descriptions of their respective strengths and weaknesses. In the current work, two analytical approaches were used: those of Thomasson and others (1960) and Theis and others (1963). These methods require data on the pumping rate, drawdown, and duration of the test to determine transmissivity, all of which are commonly found on well-driller reports.

The approach of Thomasson and others (1960) is based on the Dupuit-Theim equation and relates transmissivity to specific capacity with a dimensionless constant, C_c, which incorporates the well radius. The main assumptions of this method are that the well-performance test is not ended until the water level in the well stabilizes (reaches steady state) and that storage coefficient, partial penetration of the well into the aquifer, and well efficiency do not significantly affect results. If both specific capacity and transmissivity are simplified to units of foot squared per day (ft^2/d), Thomasson and others (1960) found Cc to average 1.2 for valley-fill sediments in California, which is the value used in the current report. Analyses for the current work assumed a value of 0.2 for the storage coefficient. This relation is expressed as:

$$T = C_c S_c, \qquad (3)$$

where

T is transmissivity in length/time units,
C_c is a constant that incorporates well radius and radius of influence, dimensionless, and
S_c is specific capacity in length/time units.

The Theis and others (1963) solution is based on the Theis nonequilibrium equation. The assumptions of this method are: a fully penetrating well; a homogeneous, isotropic aquifer (hydraulic properties are equal in all directions); negligible well loss; and that the effective radius equals the well radius (Mace, 2001). Analyses for the current work assumed a value of 0.2 for the storage coefficient. The Theis and others (1963) equation is:

$$S_c = \frac{4\pi T}{\left[\ln\left(\frac{2.25 T t_p}{r_w^2 S} \right) \right]}, \qquad (4)$$

where

S_c is specific capacity in length/time units,
T is transmissivity in length/time units,
t_p is pumping time in time units,
r_w is well radius in length units, and
S is storage coefficient (storativity), dimensionless.

Of the approximately 2,900 well-driller reports available for wells drilled in the Wood River Valley, data for only 72 wells were adequate for estimating transmissivity; data for an additional nine wells found in Smith (1959) and Frenzel (1989) were also included (appendix E). Although various revisions of the Idaho well-report form include fields for reporting pump-test data, this information is commonly absent or incomplete. In addition, "pumping depth" on some versions of the form was interpreted ambiguously by drillers and may indicate either the pumping water level or the depth at which the pump was set. Tests with pumping rates of less

than 50 gal/min, durations of less than 1 hour (hr), or that were conducted with a bailer also were excluded from the analyses.

The inclusion of well-performance tests conducted for short durations and low flow rates may violate the assumption that drawdown during the test has stabilized; however, by excluding these tests, the number of calculated transmissivity values would have been reduced significantly. For a geohydrologic framework of the Snake River Plain, Whitehead (1992) compared transmissivities estimated from multiple-well aquifer-test data to those estimated on the basis of specific-capacity values from well-performance tests and the Theis and others (1963) method. Whitehead found that the aquifer-test derived values tended to be as much as an order of magnitude greater than the specific-capacity derived values. This difference was attributed to well efficiency loss and the inclusion of non-fully penetrating wells which yielded lower specific-capacity values. In addition, specific capacities tend to decrease with pumping rate. For the current report, one previously reported transmissivity value from Smith (1959) and eight hydraulic conductivity values from Frenzel (1989) fall within the range of values estimated from specific capacity. Therefore, it is believed that the well-performance tests used in the current report provide valid estimates of transmissivity and hydraulic conductivity, yet they should be considered approximate values that are accurate to within about an order of magnitude.

Analysis of well-performance tests were conducted by programming the equations into a spreadsheet. Because the Theis and others (1963) equation can not be solved directly for transmissivity, the spreadsheet incorporates iterative calculations similar to those described in Mace (2001). The two values calculated by the different approaches were rounded to two significant figures.

Because transmissivity is the product of hydraulic conductivity times saturated thickness, transmissivity values differed significantly between tributary canyons and main valley because of differences in saturated thickness and possible boundary effects. In wells that reached bedrock, saturated thickness was determined by subtracting the water level on the well-driller report from bedrock depth. For wells that did not penetrate to bedrock, bedrock depth was determined from the bedrock surface map produced for the current report (pl. 1). If this surface was higher than the well depth, the bedrock depth was assumed to equal well depth. Saturated thickness was determined for wells exhibiting confined conditions from data included in the well-driller report and, if necessary, the bedrock surface map (pl. 1). Hydraulic conductivity is discussed in section, "Hydraulic Conductivity."

Hydrogeologic Framework

Geologic Setting

A complex series of geologic events is responsible for the distribution and hydrogeologic characteristics of water-bearing rocks in the Wood River Valley. A simplified version of this geologic history, as it pertains to groundwater, is presented here to provide a context for the following description of hydrogeologic units in the study area. A geologic time scale with these hydrogeologic units and main geologic events of the Wood River Valley area is shown in table 1, and a generalized geologic map of the Wood River Valley and surrounding area is shown in figure 4.

Much of the current understanding of the geology of the Wood River Valley and surrounding area stems from the description and understanding of mineral deposits. Minerals mined in the area include lead, silver, zinc, gold, copper, tungsten, antimony, molybdenum, bismuth, iron, and barite, with the age of mineralization ranging from Middle Devonian to Quaternary time (Worl and Johnson, 1995a). Although these deposits have little bearing on the hydrogeology of the study area, their presence, and particularly the spoils and tailings from historical mining activity, have affected water quality in localized areas.

Pre-Quaternary History

The oldest rocks in or adjacent to the study area are intensively metamorphosed Proterozoic rocks that have been age-dated to 2 billion years (Dover, 1983). Emplacement of these rocks was followed by the deposition of sedimentary rocks, which were subsequently metamorphosed, about 1.3 billion years ago (Dover, 1983). After a hiatus, Ordovician and Silurian marine sediments were deposited on a west-facing continental margin represented by the Phi Kappa and Trail Creek Formations (Rodgers and others, 1995). During Devonian time, the Milligen Formation was deposited on this continental margin as it was uplifted and deformed by tectonism leading up to the Late Devonian-Early Mississippian Antler orogeny, which then led to the deposition of the Mississippian Copper Basin Formation northeast of the study area (Hall, 1985; Link and others, 1988, 1995a; Rodgers and others, 1995). During Middle Pennsylvanian to Early Permian time, the Wood River basin developed, probably as a result of the Ancestral Rockies orogeny, into which marine sedimentary rocks of the Sun Valley Group were deposited during two phases of subsidence (Hall, 1985; Link and others, 1988; Link and others, 1995a; Rodgers and others, 1995).

During Late Jurassic to Cretaceous time, the Sevier orogeny resulted in a number of generally north-trending folds and faults in the Wood River Valley area (Rodgers and others, 1995). Because this deformation affected the lithologically similar Paleozoic sedimentary rocks of the "central Idaho black shale mineral belt" of Hall (1985), structural relations can be difficult to interpret and are often ambiguous. Although a number of regional top-to-the-east thrust faults have been mapped with displacements ranging from a few miles to 120 mi (Hall, 1985), Rodgers and others (1995) indicated that many of these thrust faults are in fact shear zones, normal or dip-slip faults, or depositional contacts.

In Late Cretaceous time, the granitic Idaho batholith was intruded into older rocks as a result of changes in the speed and direction of tectonic plates to the west (Vallier and Brooks, 1987; Kiilsgaard and others, 2001). The Idaho batholith covers approximately 15,000 mi^2, encompassing the rugged, largely roadless mountains of central Idaho (DeGrey and others, 2011). After emplacement of the Idaho batholith during the Late Cretaceous, it was exhumed by middle Eocene time when the Challis Volcanic Group erupted (Rodgers and others, 1995). Additional folding, faulting, and dike intrusions occurred before, during, and after the Challis volcanism and extended into the Oligocene.

The eastern Snake River Plain developed during the Miocene as the North American Plate moved to the southwest over a mantle plume, thus creating the Yellowstone hotspot track. As the plate migrated over the hotspot, a series of calderas were formed; from these calderas, large volumes of rhyolitic rocks of the Idavada Volcanics erupted (Link, 2011). A temporary topographic high corresponded to the location of the hotspot that subsided with crustal cooling as the plate moved to the southwest (Link and others, 2005; Beranek and others, 2006). This subsidence, in combination with probable caldera collapse and crustal loading from Quaternary basalts, is responsible for the topography of the modern eastern Snake River Plain (Hackett and Morgan, 1988; Link, 2011). The ancestral Big Wood River was captured by the Snake River drainage during the Pliocene (Beranek and others, 2006).

Miocene basin-and-range normal faulting began at about the same time as Idavada volcanism. The Wood River graben subsided between various normal faults including the north-trending Sun Valley fault zone, which runs along the eastern side of the Wood River Valley (Link and Rodgers, 1995; Rodgers and others, 1995; Breckenridge and Othberg, 2006). The Sun Valley fault cuts Quaternary deposits adjacent to Trail Creek Road, about 2 mi north of its junction with Idaho 75 (Hall and others, 1978; Rodgers and others, 1995). Elsewhere on the map of Hall and others (1978), the Sun Valley fault is more commonly mapped as covered by Quaternary sediment. Because these authors have mapped Quaternary sediments offset by the Sun Valley fault, movement along the Sun Valley fault zone may have continued sporadically to the present day, similar to other basin-and-range normal faults in the region. On April 26, 1969, an earthquake with local magnitude 4.75 and Mercalli Intensity VI cracked concrete floors in the Warm Springs drainage and Ketchum (Breckenridge and others, 1984; Idaho Geological Survey, 2012)

Table 1. Geologic time scale with hydrogeologic units of the Wood River Valley area and significant orogenies and geologic events.

[Modified from U.S. Geological Survey Geologic Names Committee (2010).]

Eonothem / Eon	Erathe / Era	System, Subsystem / Period, Subperiod	Series / Epoch	Hydrogeologic unit; Map symbol	Orogeny or event
Phanerozoic (542 Ma to present)	Cenozoic (66 Ma to present)	Quaternary (2 6 Ma to present)	Holocene (11,700 y to present)	Quaternary sedimentary deposits and basalts; Qu	
			Pleistocene (2 6 Ma to 11,700 y)		
		Tertiary (66 to 2 6 Ma)	Pliocene (5 3 to 2 6 Ma)		
			Miocene (23 to 5 3 Ma)	Idavada Volcanics; Tv	Hotspot track Basin-and-range
			Oligocene (34 to 23 Ma)		
			Eocene (56 to 34 Ma)	Challis Volcanic Group; Tcv Tertiary intrusives; Ti	
			Paleocene (66 to 56 Ma)		
	Mesozoic (251 to 66 Ma)	Cretaceous (146 to 66 Ma)	Upper/Late (100 to 66 Ma)	Idaho batholith; Kg	Idaho batholith
			Lower/Early (146 to 100 Ma)		
		Jurassic (200 to 146 Ma)	Upper/Late (161 to 146 Ma)		Sevier orogeny
			Middle (176 to 161 Ma)		
			Lower/Early (200 Ma to 176 Ma)		
		Triassic (251 to 200 Ma)	Upper/Late (229 to 200 Ma)		
			Middle (245 to 229 Ma)		
			Lower/Early (251 to 245 Ma)		
	Paleozoic (542 to 251 Ma)	Permian (299 to 251 Ma)			
		Pennsylvanian (318 to 299 Ma)	Upper/Late (307 to 299 Ma)		
			Middle (312 to 307 Ma)	Sun Valley Group; Psz	Ancestral Rockies orogeny
			Lower/Early (318 to 312 Ma)		
		Mississippian (359 to 318 Ma)	Upper/Late (328 to 318 Ma)		
			Middle (345 to 328 Ma)	Copper Basin Formation; Mcb	
			Lower/Early (359 to 345 Ma)		
		Devonian (416 to 359 Ma)	Upper/Late (385 to 359 Ma)	Milligen Formation; Dm	Antler orogeny
			Middle (398 to 385 Ma)		
			Lower/Early (416 to 398 Ma)		
		Silurian (444 to 416 Ma)		Trail Creek Formation; YS	
		Ordivician (488 to 444 Ma)		Phi Kappa Formation; YS	
		Cambrian (542 to 488 Ma)			
Proterozoic (2,500 to 542 Ma)	Neoproterozoic (1,000 to 542 Ma)			Undifferentiated; YS	
	Mesoproterozoic (1,600 to 1,000 Ma)			Undifferentiated; YS	
	Paleoproterozoic (2,500 to 1,600 Ma)			Undifferentiated; YS	

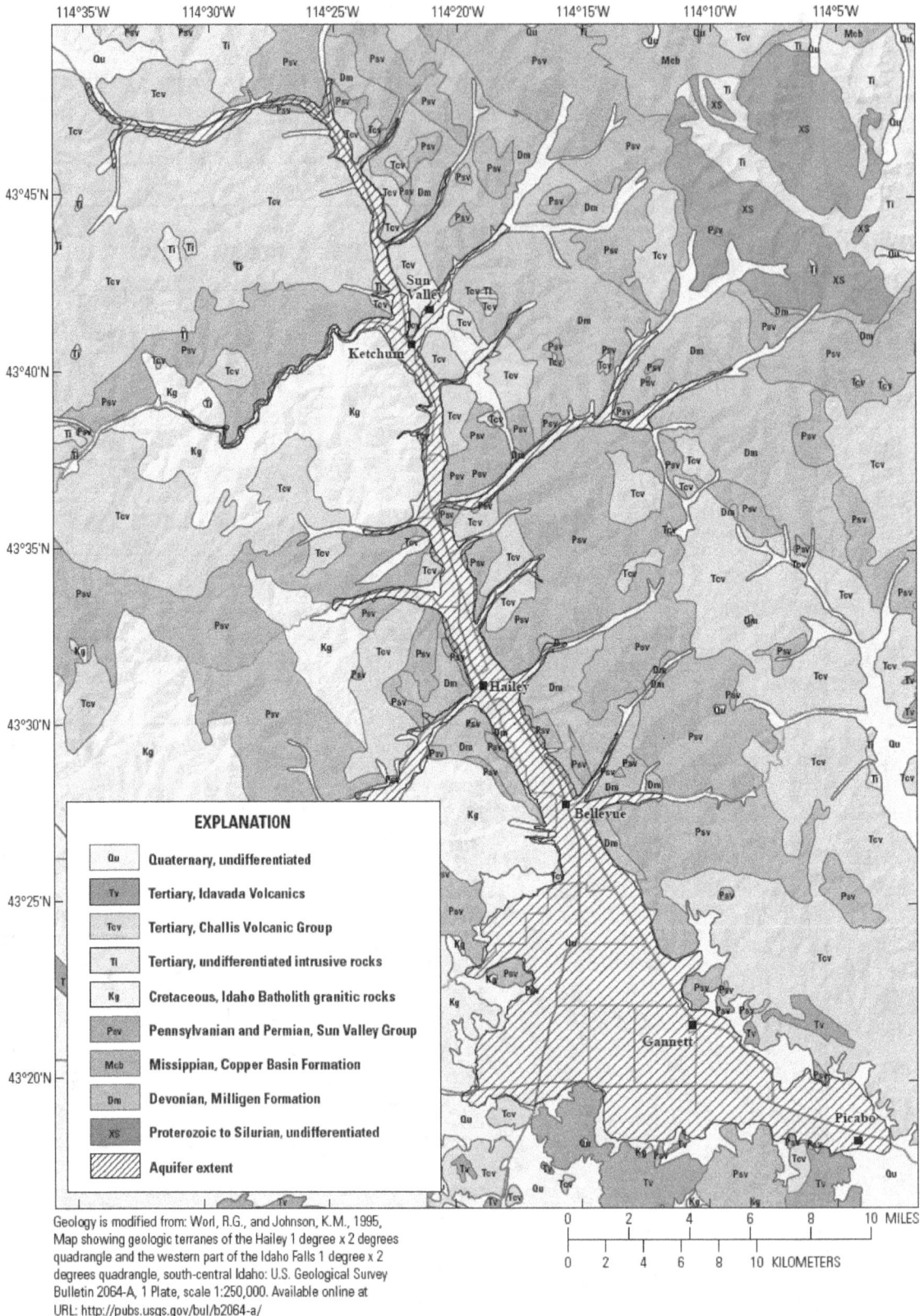

Geology is modified from: Worl, R.G., and Johnson, K.M., 1995,
Map showing geologic terranes of the Hailey 1 degree x 2 degrees
quadrangle and the western part of the Idaho Falls 1 degree x 2
degrees quadrangle, south-central Idaho: U.S. Geological Survey
Bulletin 2064-A, 1 Plate, scale 1:250,000. Available online at
URL: http://pubs.usgs.gov/bul/b2064-a/

Figure 4.   Generalized geologic map of the Wood River Valley and surrounding area, south-central Idaho. Modified from the geologic terrane map of Worl and Johnson (1995b). Because the terrane map emphasizes rock age and lithology, geologic units are not differentiated. Consequently, this map uses the predominant geologic unit present within the mapping unit as the unit name. The reader is referred to Worl and others (1991) for a true geologic map of the entire Wood River Valley and surrounding area.

Quaternary History

Geologic events during the Quaternary Period are responsible for the alluvial sediments and basalts that form the bulk of the Wood River Valley aquifer system. Schmidt (1962) described a sequence of deposition and erosion by the Big Wood River, basalt flows in the Picabo and Stanton Crossing areas, and glaciation in the upper Big Wood River drainage and tributary canyons. While Schmidt's work is considered fundamentally correct in describing a number of cycles of alluvial and lacustrine deposition primarily controlled by Quaternary volcanism, recent radiometric dating of various lava flows contradict parts of his proposed chronology and has led to revision of some his stratigraphy (see section, "Basalts and Associated Deposits").

In early Pleistocene time, the ancestral Big Wood River flowed onto a broad alluvial fan with its apex at Bellevue (the Bellevue fan) and exited the valley either to the east or west of the Timmerman and Picabo Hills. A series of lava flows then erupted, which dammed and diverted the river between the eastern and western sides of the Timmerman and Picabo Hills, possibly multiple times, resulting in a sequence of fluvial and lacustrine sediments of some thickness on the Bellevue fan that interfinger with basalts to the southeast and possibly southwest. If basin-and-range faulting continued into the Quaternary it is likely that these faults also influenced drainage of the ancestral Big Wood River and deposition of the Bellevue fan.

Adding to the complexity of the Quaternary depositional history are at least two episodes of glaciation preserved in the landscapes of the Big Wood River drainage (Schmidt, 1962; Pearce and others, 1988). Although these glaciers were confined to the surrounding mountains and upper reaches of tributary canyons, and no ice lobe advanced down the main valley, they furnished the Big Wood River with a source of sediment ranging in size from clay to boulders (Pearce and others, 1988). The two glacial episodes created glacial outwash that contributed to stream aggradation in the main valley (including the Bellevue fan) and tributaries. The Wood River Valley assumed its current form when the Big Wood River incised about 30 ft, probably from a combination of uplift and episodic climatic changes (Schmidt, 1962).

Hydrogeologic Units

The primary goal of this report is the characterization of the Wood River Valley aquifer system, which, as noted above, is mostly composed of Quaternary-age sediments and basalts of the Wood River Valley and its tributaries. The older igneous, sedimentary, or metamorphic rocks that underlie these sediments, commonly referred to as bedrock, are used for groundwater supply, although it is unclear to what extent these rocks are hydraulically connected to the Quaternary-age sediments and basalts that form the bulk of the Wood River

Valley aquifer system. Bartolino (2009) did not consider them a significant source of groundwater inflow to the Quaternary alluvium.

Paleozoic Sedimentary Rocks

The oldest hydrogeologically significant rocks that surround and underlie the Wood River Valley aquifer system are Ordovician- to Permian-age rocks of the "central Idaho black shale mineral belt" of Hall (1985). These marine sedimentary rocks are "predominantly black, fine-grained, siliceous argillite...limy sandstone, shale, and siltstone" (Hall, 1985). Most of these rocks are poorly exposed, contain few fossils, and have been folded, faulted, and intruded multiple times. As a result, mapping and correlating these units is problematic, and definitions and stratigraphic assignments have been disputed in the literature.

Ordovician and Silurian Rocks

Ordovician and Silurian rocks of the Phi Kappa and Trail Creek Formations and other undifferentiated units are found in the headwaters of Trail Creek and the East Fork Wood River (Worl and others, 1991; Link and others, 1995a, 1995b) (figs. 1 and 4). Wust and Link (1988) reported another outcrop of Phi Kappa Formation between Trail Creek Road and Trail Creek about 8.2 mi east of the junction of Idaho 75 and Sun Valley Road; currently (2012), the outcrop appears to be covered by loose rock from road construction. These weakly metamorphosed, heavily faulted rocks include massive, gray, fine-grained quartzite; red-weathering shale, argillite, and thin limestone; and buff-weathering metasiltstone and quartzite (Link and others, 1995a). Although springs may issue from fractures in these rocks and contribute water to Trail Creek, the locations of mapped outcrops make it unlikely that they supply water to wells in the study area.

Milligen Formation (Upper to Lower Devonian)

The Upper to Lower Devonian Milligen Formation is exposed primarily on slopes of the eastern Wood River Valley and eastern tributary canyons, but also in the Croy Creek drainage and the western side of the main valley north and south of Hailey (Worl and others, 1991) (figs. 1 and 4). The Milligen Formation is approximately 4,000 ft thick and was divided into a number of informal members by Turner and Otto (1995). Rock types include dark-gray to black argillite, siltstone, micritic limestone, dolomitic siltstone, diamictite, chert, sandstone, quartzite, limestone, and conglomerate (Hall, 1985; Turner and Otto, 1995). The Milligen Formation underlies Quaternary alluvium in tributary canyons (and possibly the main valley), and many wells are completed in it. Horizontal hydraulic conductivity possibly ranges over 12 orders of magnitude: values for similar rock types taken from the literature range from 3×10^{-8} to 2,000 ft/d; the larger values are for fractured rock (Spitz and Moreno, 1996).

Sun Valley Group

Mahoney and others (1991) defined the Sun Valley Group to include the lithologically similar Wood River, Dollarhide, and Grand Prize Formations (the Grand Prize Formation crops out northwest of the study area and is not discussed here).

Wood River Formation (Middle Pennsylvanian to Early Permian)

The Middle Pennsylvanian to Early Permian Wood River Formation is exposed primarily on slopes of the eastern Wood River Valley and tributary canyons, as well as in the western Wood River Valley and tributary canyons north of Deer Creek (Worl and others, 1991) (figs. 1 and 4). The Wood River Formation is about 10,000 ft thick (Hall, 1985) and was divided into three formal members by Mahoney and others (1991). Lithology is siliceous conglomerate overlain by sequences of light gray limy sandstone, fine-grained quartzite, and silty limestone (Hall, 1985). These rocks provide water to wells in the study area. Horizontal hydraulic conductivity likely ranges over 10 orders of magnitude: values for similar rock types taken from the literature range from 1×10^{-6} to 2,000 ft/d; the larger values are for fractured rock (Spitz and Moreno, 1996).

Dollarhide Formation (Permian)

The Permian Dollarhide Formation is exposed in a northwest trending band from Poverty Flat to near Dollarhide Summit at the head of the Warm Springs Creek drainage (Worl and others, 1991) (figs. 1 and 4). The Dollarhide Formation is about 6,500 ft thick, and various authors have recognized three informal members (Link and others, 1995a). Lithology is a sequence of "interbedded dark-gray carbonaceous limestone, [siltstone], fine-grained quartzite, sandstone, and minor granule conglomerate" (Hall, 1985). Outcrops of the Dollarhide Formation were previously mapped as Wood River Formation although the former is typically darker and finer grained than the latter (Hall, 1985). These rocks provide water to wells in the study area. Horizontal hydraulic conductivity likely ranges over 10 orders of magnitude with values in the literature ranging from 1×10^{-6} to 2,000 ft/d; the larger values are for fractured rock (Spitz and Moreno, 1996).

Cretaceous Intrusive Rocks

Granitic intrusive rocks of the Atlanta lobe of the Late Cretaceous Idaho batholith are exposed on the western side of the Wood River Valley in a northwest-trending band of discontinuous outcrops from about Stanton Crossing to Warm Springs Creek (Johnson and others, 1988; Worl and others, 1991) (figs. 1 and 4). Lewis (1989) and Kiilsgaard and others (2001) divided the batholithic rocks of the area into three groups: quartz diorite, a potassium-rich suite of hornblende-biotite granodiorite and granite, and a sodium-rich suite of biotite granodiorite and hornblende-biotite

granodiorite. Each of these rock groups are enriched in different minerals and elements with varying water-quality implications. These rocks likely provide water to wells on the western side of the Bellevue fan and the Deer Creek, Croy Creek, and Warm Springs Creek drainages. Granitic rocks have virtually no primary porosity, only secondary porosity due to fractures. Spitz and Moreno (1996) give values of hydraulic conductivity ranging from 1×10^{-6} to 7×10^{-6} ft/d for granite, 0.9 to 15 ft/d for weathered granite, and 3×10^{-2} to 300 ft/d for fractured granite.

Tertiary Igneous Rocks

Some of the most dramatic rock outcrops in the Wood River Valley are exposures of Tertiary-age volcanic and intrusive rocks. These rocks form the eastern side of the valley in the vicinity of Elkhorn Gulch and both sides of the valley above Ketchum as well as most of the western face of the Boulder Mountains.

Challis Volcanic Group (Eocene)

The two largest exposures of the Eocene Challis Volcanic Group in the study area are on the west side of the Wood River Valley north of Ketchum and the east side of the Bellevue fan; however, smaller outcrops are found in many tributary canyons of the Wood River Valley (Worl and others, 1991) (figs. 1 and 4). Thickness of the Challis Volcanic Group rocks in the study area is greater than 3,400 ft as measured by Sanford (2005) near Muldoon Summit at the head of Seaman's Gulch. Three main rock types of the Challis Volcanic Group crop out in the vicinity of the study area: volcaniclastic sedimentary rocks, dacite lava flows and flow breccia, and andesite lava flows and tuff breccia (Moye and others, 1988; Worl and others, 1991; Sanford, 2005). Geologic mapping of the area south of Baseline Road by Kauffman and Othberg (2007) and Garwood and others (2010, 2011) indicate that these rocks are "mostly medium to dark gray, pink, or purple hornblende dacite porphyry." Given their large areal extent in the Wood River Valley and its tributaries, rocks of the Challis Volcanic Group probably yield water to wells in the Elkhorn, Adams, and Oregon Gulch areas as well as in the Fox, Lake, and Eagle Creek areas. Hydraulic conductivity probably ranges over 15 orders of magnitude, from about 5.3×10^{-11} to 5.1×10^{4} ft/d (Spitz and Moreno, 1996); larger values are probably due to fracturing or weathering.

A few stocks or dikes of a gray, porphyritic dacite are exposed in the upper Warm Springs drainage and the area between Adams Gulch and Fox Creek (figs. 1 and 4); these rocks were mapped as Eocene in age by Worl and others (1991). Because these shallow intrusive rocks cross-cut rocks of both the Cretaceous-age Idaho batholith and Eocene Challis Volcanics, they probably correspond to the final stage of Challis volcanism described by Moye and others (1988) as well as the dacite intrusion-flow-dome complex unit described by Sanford

(2005). These rocks probably have similar values of hydraulic conductivity to other Challis Volcanic rocks. Where they cut more permeable rocks they may serve as barriers to flow.

Idavada Volcanics (Miocene)

Outcrops of the Miocene Idavada Volcanics are found in the Timmerman Hills, Picabo Hills, and the area southeast of Gannett (Worl and others, 1991) (figs. 1 and 4). These rocks were mapped as Picabo Tuff by Schmidt (1962), Kauffman and Othberg (2007), and Garwood and others (2010, 2011). Garwood and others (2010) describe these rocks as "light gray, tan, and purplish tan crystal-poor rhyolite tuff" although Schmidt (1962) also included interbedded tuffaceous sediments. Sanford (2005) indicated that the Idavada Volcanics are up to 700 ft thick in the Little Wood River drainage immediately east of the study area. Although these rocks have primary porosity due to vesicles and intergranular voids, they are likely not a source of water to wells or springs except possibly in the Timmerman and Picabo Hills area where a number of springs are mapped. Values of hydraulic conductivity may range from about 4.0×10^{-5} to 130 ft/d (Spitz and Moreno, 1996).

Quaternary Sedimentary Deposits and Basalt

Quaternary-age sediments and basalt are the primary sources of groundwater in the Wood River Valley aquifer system. As described in section, "Quaternary History," a sequence of deposition and erosion by the Big Wood River, basalt flows in the Picabo and Stanton Crossing areas, and glaciation in the upper Big Wood River drainage and tributary canyons created the bulk of the Wood River Valley aquifer system. In addition to two basalt units, Schmidt (1962) mapped and informally named 12 units of unconsolidated Quaternary sediment in the Bellevue area; these units fall into three main groups: deposits impounded by lava dams, deposits related to glacial activity, and deposits formed by erosional processes. Subsequent authors (such as Moreland, 1977, and Brockway and Kahlown, 1994) primarily concerned with defining hydrostratigraphic units within the aquifer system have tended to divide the alluvial material into a coarse-grained sand and gravel unit, a fine-grained silt and clay unit, and a single basalt unit. This fairly simple textural/lithologic division adequately describes the aquifer system, and difficulties associated with the use of well-driller reports hinder further subdivision.

Coarse-Grained Sedimentary Deposits

The coarse-grained sediment of the Wood River Valley aquifer system is composed of rounded sand and gravel and is usually poorly sorted. Most of these coarse-grained deposits are of alluvial origin and resulted from the deposition of glacially derived sediment by the Big Wood River and its tributary streams. Because these larger sediment sizes require greater streamflow for transport, they typically represent deposition in the main stream channel. As the Big Wood River and tributary streams shifted and meandered across valley bottoms and the Bellevue fan, this alluvial sediment was deposited, often eroded, and then redeposited. Consequently, the coarse-grained deposits tend to be horizontally and vertically discontinuous (fig. 5).

A relatively minor portion of the coarse-grained deposits represent active erosional processes by colluvial transport of sediment down slopes and across pediments, as described by Smith (1962). In general, these surficial sedimentary deposits are thin and do not serve as a source of water to wells, they may, however, play a role in transporting rainfall and snowmelt to the Wood River Valley aquifer system.

Most of the water produced from the Wood River Valley aquifer system is from the coarse-grained deposits because of their greater productivity due to larger values of hydraulic conductivity. Estimated hydraulic conductivity values are discussed in section, "Hydraulic Conductivity;" the estimated values tend to represent flow from coarse-grained intervals rather than fine-grained intervals.

Fine-Grained Sedimentary Deposits

Fine-grained sediment is found in most areas of the Wood River Valley aquifer system and is either lake sediment deposited as a result of damming of the Big Wood River by lava flows or as overbank and flood deposits of the Big Wood River and its tributary streams. These fine-grained deposits commonly act as confining units of limited areal extent but also form the more areally extensive confining unit underlying the southern Bellevue fan.

Previous workers have described a confining unit that separates the unconfined aquifer from the underlying confined aquifer in the area south of about Baseline Road (fig. 1). Moreland (1977) described the unit as dipping to the south, about 120 ft thick, and generally lying at a depth of about 150 ft. For the current study, well-driller reports were evaluated in order to construct hydrogeologic sections (fig. 5), and a map of the estimated altitude of the top and corresponding thickness of the uppermost unit of fine-grained sediment (fig. 6) was prepared (as described in the section, "Compilation of Fine-Grained Sediment Map"). The hydrogeologic section A-A' in figure 5 generally agrees with Moreland's (1977) section B-B' and shows fine-grained sediment of the confining unit extending and thinning over Quaternary basalt, thus allowing the coarse-grained sediment of the unconfined and confined aquifers to hydraulically reconnect in the area south of Gannett. Figure 6 confirms the general dip of the top of the fine-grained sediment to the south and southeast but also shows significant variability in the thickness of this unit. In general, the thickness of the uppermost fine-grained sediment unit increases to the south and decreases where it overlies Quaternary basalt.

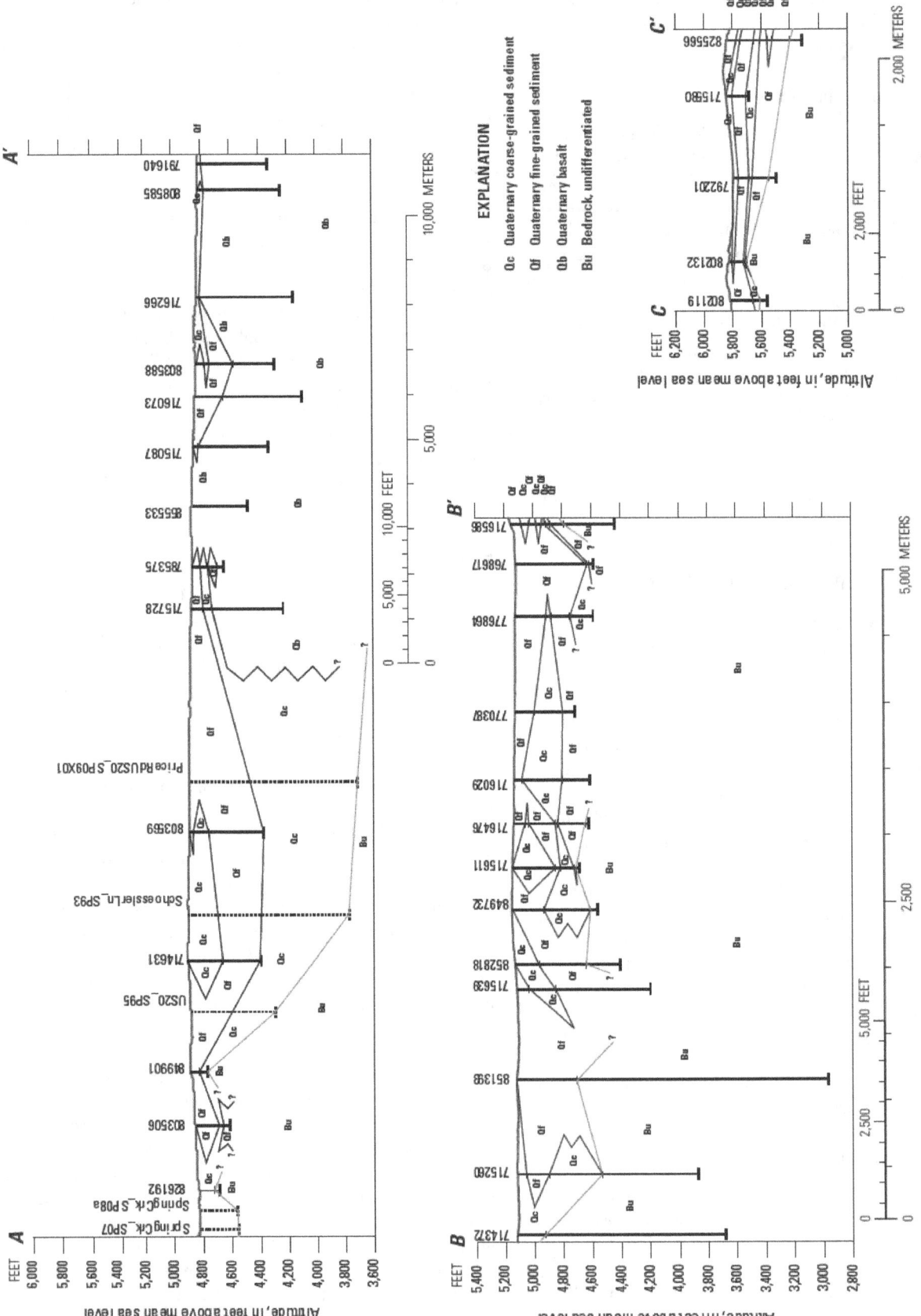

Figure 5. Hydrogeologic sections showing generalized lithologic units. Section *A–A'* trends west to east in the vicinity of U.S. 20, section *B–B'* trends west to east in the vicinity of Glendale and Pero Roads, section *C–C'* trends northwest to southeast through Ketchum. Lines of section are shown on plate 1.

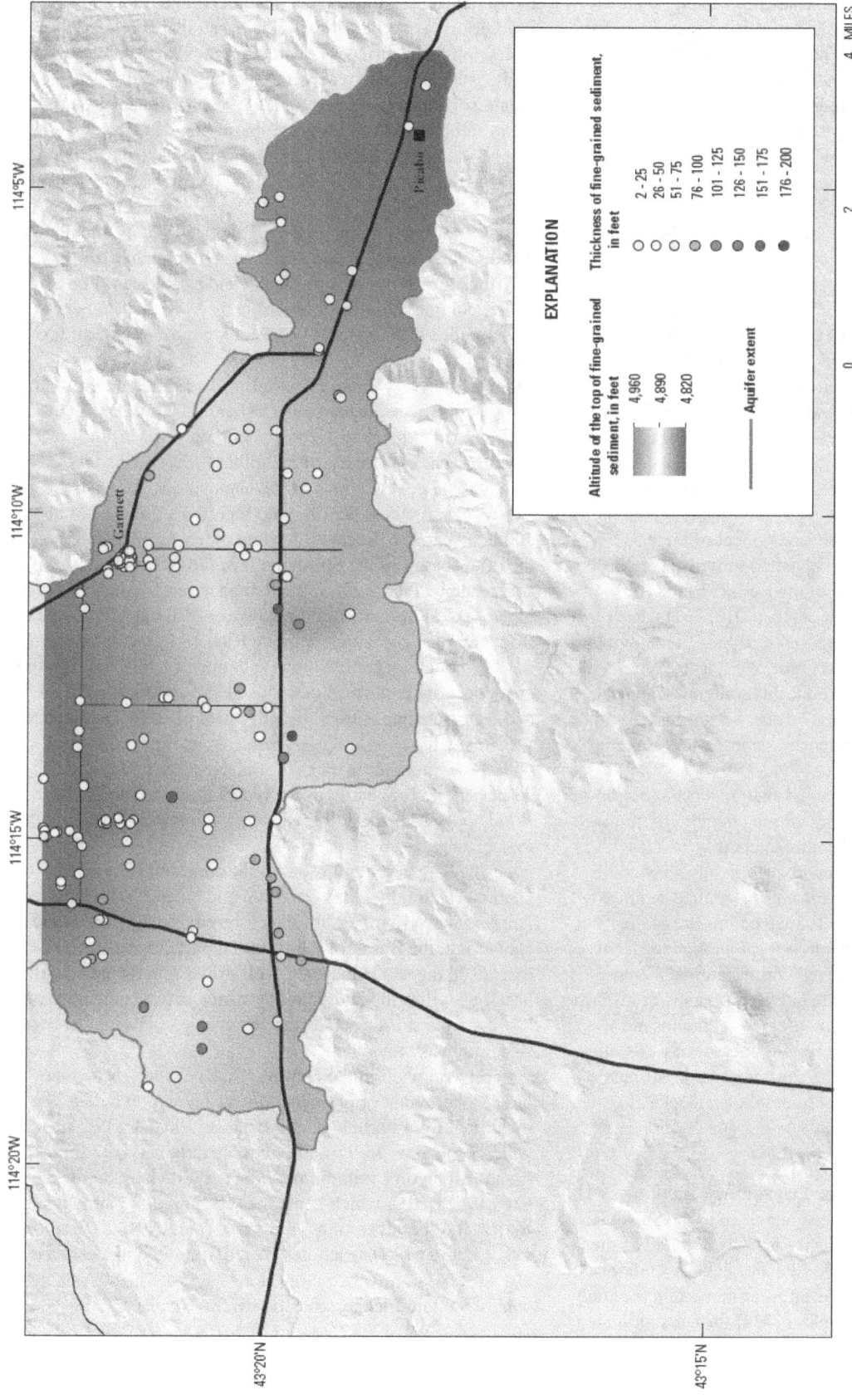

Figure 6.   Estimated altitude of the top and corresponding thickness of the uppermost unit of fine-grained sediment within the Wood River Valley aquifer system, southern Wood River Valley, south-central Idaho.

Most fine-grained sediment elsewhere in the Wood River Valley aquifer system is glacially derived and was deposited along the Big Wood River and its tributaries at higher stream stages as clay and silt settled out of slower moving water on floodplains. As with the coarse-grained deposits, repeated deposition and erosion as the streams migrated across the main valley and tributary floors caused these fine-grained deposits to be relatively thin and discontinuous in the upper valley (fig. 5, hydrogeologic section C-C').

As described in the lithologic section of well-driller reports, fine-grained deposits may be recorded solely as clay; more typically they are recorded as a combination of textural sizes including clay, such as sand-gravel-clay, clayey gravel, or sandy clay. Although color is not always noted on well-driller reports, recorded colors include brown, tan, red, pink, yellow, white, blue, gray, and black. Brown is the most common color; blue and black clay commonly is used as a marker by local drillers. A few well-driller reports describe caliche or hardpan within the Quaternary sedimentary deposits, which likely represents soil formation during periods of geomorphic stability.

Moreland (1977) described several feet of loess, predominantly silt-sized wind-deposited sediment, as underlying parts of the surface of the Bellevue fan. The geologic maps of Garwood and others (2010, 2011) describe various basalts as covered by up to 6 ft of loess, although the loess is not mapped as a separate unit. Well-driller reports typically record several feet of soil; this soil may represent discontinuous exposures of loess or may be soil formed in the coarse- and fine-grained sedimentary deposits that make up the fan.

Where deposits of fine-grained sediment act as confining units separating zones exhibiting varying degrees of confined hydraulic head, the degree of confinement may be difficult to evaluate because well-driller reports commonly show wells screened across multiple fine- and coarse-grained sediment zones. Water levels in such wells are a composite of the different hydraulic heads and may appear anomalous from levels in adjacent wells. Such well construction also may allow flow through the well bore between zones with differing hydraulic head. Where these fine-grained sediments are saturated and included in the screened interval of wells, they yield less water to wells than the coarse-grained sediments because of their lower values of hydraulic conductivity.

Basalts and Associated Deposits

Schmidt (1962) divided the Quaternary basalt flows of his study area into two units and assigned them to the Snake River basalt of Stearns (1938), which was later reclassified as the Snake River Group by Malde and Powers (1962). Only one of these two units, the Bellevue basalt, is found within the study area of the current report. Schmidt (1962) further subdivided the Bellevue basalt into seven members. The Bellevue basalt is described as porphyritic olivine basalt that is primarily black to gray on fresh surfaces (weathering to brown) and is mostly of the pahoehoe flow type with interflow zones of brecciated basalt, soil, and cinders (Schmidt, 1962). This sequence of flows and interflow zones is similar to that described by Hughes and others (1999) for the eastern Snake River Plain. Extent of the basalt in the Wood River Valley aquifer system is shown on plate 1.

Recent work by the Idaho Geological Survey, including detailed field mapping and radiometric-dating, has led to revisions to Schmidt's stratigraphy and chronology. For the hydrogeologic framework described here, three members of Schmidt's (1962) Bellevue basalt are most important for the current hydrogeologic framework. The first, the Wind Ridge member of Schmidt (1962), was renamed the Basalt of Wind Ridge by Kauffman and Othberg (2007), who reported flow ages ranging from 1.1 to 1.5 million years for different outcrops. The Hay basalt of Schmidt (1962) retained its name on the geologic map of the Gannett quadrangle by Garwood and others (2010), who reported a flow age of about 500,000 years by Breckenridge and others (2003). Finally, the Priest basalt of the Shoshone lava flow (Schmidt, 1962) was renamed the Basalt of the Picabo desert by Garwood and others (2011); they reported a flow age of about 460,000 years from a written communication by Richard Esser, dated 2005.

Although recent radiometric dating of basalt flows in the area generally confirms Schmidt's (1962) cycles of alluvial and lacustrine deposition primarily controlled by Quaternary volcanism, his chronology of basalt flows is not consistent with these radiometric dates. For instance, Schmidt (1962) described the Hay basalt as the oldest flow and the Basalt of Wind Ridge as next oldest, whereas radiometric dates suggest the opposite.

The Hay basalt crops out in the area south and east of Gannett (Schmidt, 1962; Garwood and others, 2010, 2011). Approximately 2 mi east of the easternmost exposure of the Hay basalt, the Basalt of the Picabo desert crops out and extends to the south and east. Well-driller reports indicate that basalt underlies the alluvium in the area between the outcrops; however, the subsurface extent and relation between the two basalt units is uncertain.

Several wells at the Hayspur Fish Hatchery penetrate basalt to alluvium or probable Sun Valley Group rocks, including, from north to south: well 715300 (01S 19E 13BD), which penetrates 75 ft of alluvium, underlain by 137 ft of basalt resting on a minimum of 76 ft of alluvium; 853072 (01S 19E 13DB), which penetrates 75 ft of alluvium underlain by 108 ft of basalt resting on limestone; and 791629 (01S 19E 13DC), which penetrates 74 ft of alluvium, underlain by 101 ft of basalt resting on limestone (fig. 1). The reasonably consistent unit thicknesses between these wells and the presence of alluvium beneath basalt in the center well, 853072 (01S 19E 13DB), suggests the presence of a paleo-stream incised into Sun Valley Group rocks, which were subsequently overlain by basalt flows and capped by alluvium of the present Silver Creek.

The subsurface extent of basalt in the study area is confined to the southeastern Bellevue fan (pl. 1) and only one report for a well completed in basalt was suitable for estimating hydraulic conductivity (pl. 1; appendix E). Well 791635 (01S 19E 20CD) is open to 102 ft of basalt and sedimentary interbeds and 29 ft of Sun Valley Group limestone; it is discussed in more detail in section, "Hydraulic Conductivity."

Consolidated Rock Surface Underlying Quaternary Sediment and Thickness of Quaternary Sediment

A map of the approximate altitude of the consolidated rock surface formed by pre-Quaternary bedrock and Quaternary basalt is shown on plate 1. This bedrock surface represents the base of the Quaternary sediment that constitutes the bulk of the Wood River Valley aquifer system and was compiled as described in the section, "Study Methods." The top of buried Quaternary basalt is included as part of the surface because few wells completely penetrate the basalt and because HVSR measurements detect the uppermost contact between sediment and basalt. The altitude of the consolidated rock surface generally mimics the land surface and decreases down tributary canyons and the main Wood River Valley; altitude ranges from more than 6,700 ft in the Baker Creek drainage to less than 4,600 ft in the south-central Bellevue fan (pl. 1).

Beneath the southeastern part of the Bellevue fan, the consolidated rock surface represents the tops of the Hay basalt and Basalt of the Picabo desert. On plate 1, it is expressed as an area greater than 4,800 ft in altitude that drops below 4,800 ft to the east along the course of Silver Creek.

There is no apparent subsurface expression of features related to basin-and-range normal faulting shown on plate 1. Similarly, there is no indication of a subsurface bedrock "reef" extending southeast from an outcrop of Challis Volcanics near the Big Wood River bridge on Glendale Road as hypothesized by Smith (1959, pl. 5).

Much of the south-central portion of the Bellevue fan is underlain by an apparent topographically closed area on the bedrock surface that appears to drain to the southwest towards Stanton Crossing (pl. 1). This low area represents the pre-Hay surface of Schmidt (1962); to the east, the Hay basalt flow dammed the drainage outlet beneath current day Silver Creek. This damming forced drainage to the western outlet at Stanton Crossing beneath the current day Big Wood River, as suggested by the lower altitude bedrock surface

beneath the southwestern Bellevue fan (fig. 1; pl. 1). This low area is primarily delineated by HVSR measurements and constrained by well-driller reports for boreholes completed in alluvium and not penetrating to bedrock. Although many of these readings were taken under windy conditions, which may have caused an overestimation of bedrock depth (Lane and others, 2008), there is consistency between measurements made with different instruments at different times and dates. Although few wells in the Bellevue fan penetrate to depths approaching the estimated surface, well 853057 (01S 19E 06 BB) was drilled 287 ft into "gravel and clay" and well 716632 (01S 19E 18 DC) was drilled to 300 ft into "gray shale" that probably indicates Challis or Idavada Volcanics (figs. 1 and 4). This topographically low area on the bedrock surface is mimicked on Moreland's map of transmissivity (1977, fig. 3), although his preferred explanation was a deeper confined system under higher hydraulic head rather than a buried incised stream channel.

Two lines of evidence indicate that granitic rocks of the Idaho batholith may underlie Quaternary sediment on the western side of the Bellevue fan: the presence of outcrops of Idaho batholith rocks in the Stanton Crossing area, the Picabo Hills, and the northern side of Poverty Flat; and the subsurface extension of a northwest-trending normal fault (downthrown to the east) that Link and Rodgers (1995) terminated beneath Glendale Road 0.25 mi east of its intersection with Idaho 75 (fig. 1). Similarly, on the basis of outcrops in the Picabo Hills and on the eastern side of the Bellevue fan and well-driller reports, Quaternary sediment and basalt on the eastern side of the Bellevue fan are probably underlain by Sun Valley Group sedimentary rocks. Locally, the Sun Valley Group may be overlain by Challis or Idavada Volcanics; alternatively, the volcanic rocks may have once been present but were removed by Quaternary erosion.

Bedrock surfaces in the tributary canyons generally mimic the land surface, with smooth transitions to the main valley. The absence of hanging valleys is consistent with the history and extent of glaciation described by previous authors and summarized in section, "Quaternary History."

The estimated thickness of Quaternary sediment comprising the bulk of the Wood River Valley aquifer system is shown in figure 7. Thickness ranges from less than a foot on main and tributary valley margins to about 350 ft in the central Bellevue fan. Because it represents the altitude of bedrock surface of plate 1 subtracted from land surface altitude, features generally mimic those of the bedrock surface. In several areas (including the south side of Croy Canyon and the main valley above Ketchum), greater thickness is indicated along valley margins. These sediments mostly represent remnants of older stream terraces left as modern streams have incised into the Pleistocene deposits.

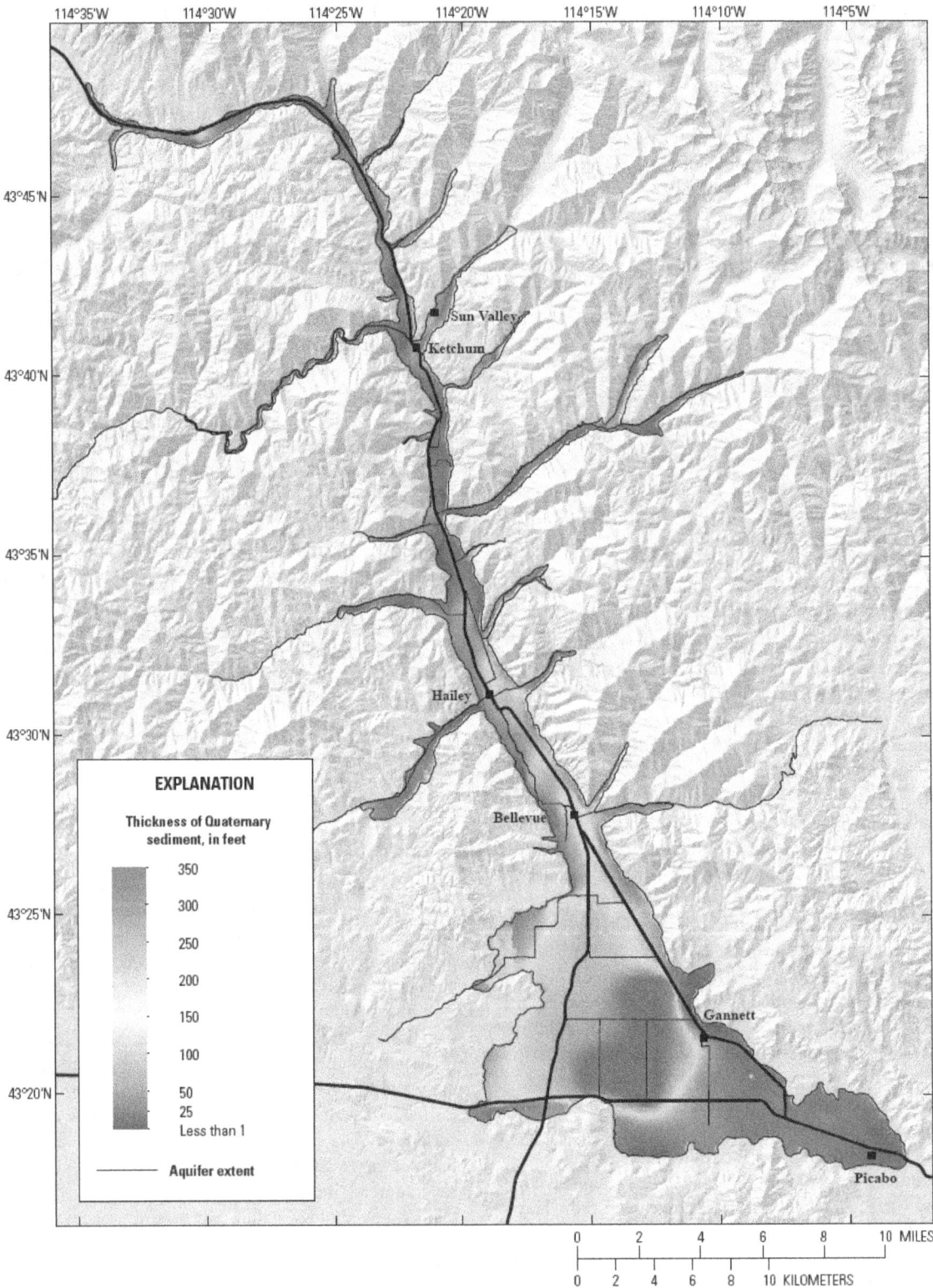

Figure 7. Estimated thickness of Quaternary sediment in the Wood River Valley aquifer system, Wood River Valley, south-central Idaho.

Hydraulic Conductivity

A map showing estimated hydraulic conductivity for 81 wells in the study area (fig. 8) was compiled from well-driller report pumping tests as described in the section, "Study Methods." Although figure 8 shows estimated hydraulic conductivity values calculated from transmissivity values estimated by the method of Theis and others (1963), Appendix E includes transmissivity values calculated by both the Thomasson and others (1960) and Theis and others (1963) methods, and the resulting hydraulic conductivity values. The map and table also include reported values from Smith (1959, p. 35) and Frenzel (1989, p. 18-19). Because the large range in aquifer thickness in the Wood River Valley aquifer system causes proportional variation in transmissivity values, transmissivity values were converted to hydraulic conductivity values because they are independent of saturated thickness. Estimated hydraulic conductivity in alluvium ranges from 1,900 ft/d along Warm Springs Creek to less than 1 ft/d in upper Croy Canyon. Of the 81 hydraulic conductivity values calculated using Theis and others (1963) transmissivity values, 2 percent were less than 1 ft/d, 9 percent were between 1 and 10 ft/d, 32 percent were between 10 and 50 ft/d, 15 percent were between 50 and 100 ft/d, 28 percent were between 100 and 500 ft/d, 6 percent were between 500 and 1,000 ft/d, and 7 percent were greater than 1,000 ft/d. Data from a well completed in bedrock (probably either Challis Volcanics or Milligen Formation) gave an estimated hydraulic conductivity value of 10 ft/d, data from one well completed in basalt (probably of Tertiary-age Idavada and (or) Challis Volcanics) gave a value of 50 ft/d, and data from three wells completed in the confined system gave values ranging from 32 to 52 ft/d (fig. 4, table 2, and appendix E).

The first published transmissivity values for the Wood River Valley aquifer system were those of Smith (1959, p. 35), who conducted five multiple-well aquifer tests, although he chose to publish values for only two of the tests (one of which is outside the aquifer boundary used in the current report) because of non-ideal well construction or uncertainty about hydrogeologic conditions. Moreland's map of transmissivity (1977, fig. 3) includes transmissivity values for all five of Smith's multiple-well aquifer tests. Figure 8 in the current report includes only Smith's (1959) single reported value of transmissivity (converted to hydraulic conductivity) within the current aquifer boundary as well as eight values of hydraulic conductivity from Frenzel (1989, p. 18–19).

Published values of hydraulic conductivity for unconsolidated sediment and basalt similar to those that constitute the Wood River Valley aquifer system are shown in table 3. Estimated hydraulic conductivity values shown in figure 8 and appendix E fall within these published ranges.

Ackerman and others (2006) described three hydrogeologic units defined for a groundwater-flow model of the Idaho National Laboratory area of the eastern Snake River Plain. Hydrogeologic units 1 and 2 are composed of basalt flows similar to those of the Hay basalt and Basalt of the Picabo desert. Hydrogeologic unit 1 (thin, densely fractured basalt), constitutes 17 basalt-flow groups, 4 andesite-flow groups, and 25 sedimentary interbeds. Hydrogeologic unit 2 (massive, less densely fractured basalt) consists of 7 basalt-flow groups, and 10 sedimentary interbeds. Results of single-well aquifer tests in 67 wells completed in hydrogeologic unit 1 indicated that the hydraulic conductivity of these rocks ranges from about 0.01 to 24,000 ft/d; in hydrogeologic unit 2, single-well aquifer tests in four wells yielded hydraulic conductivity estimates ranging from 6.5 to 14,000 ft/d. The lone hydraulic conductivity value for a well completed in basalt for the current study falls within this range.

Groundwater Movement

Groundwater movement through an aquifer system is the result of water flowing from areas of higher hydraulic head to areas of lower hydraulic head; such head distributions are described by water-table maps for an unconfined aquifer and potentiometric-surface maps for a confined aquifer. The most recently compiled water-table and potentiometric-surface maps of the Wood River Valley aquifer system are those of Skinner and others (2007). A groundwater budget contributes to the understanding of hydraulic head by identifying and quantifying the recharge to and discharge from the aquifer system; the most recently published budget is that of Bartolino (2009). However, the hydrogeology of the aquifer system also controls the distribution of hydraulic head and location and amount of recharge and discharge.

In general, the pattern of groundwater movement through the Wood River Valley aquifer system is relatively straightforward, and previous authors have agreed that groundwater under unconfined conditions moves down valley to the Bellevue fan, where it either enters a deeper confined aquifer or remains in a shallow unconfined aquifer; as the confining unit thins and fingers out the two aquifers appear to hydraulically reconnect in the area south of Gannett (fig. 1). Recharge is primarily from precipitation or the infiltration of streamflow (in losing reaches) and natural discharge is primarily through springs and seeps to streams or outflow from the aquifer system. Rerouting of surface water into a network of irrigation canals in the late 19th century, construction of water wells, and increased demand have affected groundwater flow, but the overall direction of groundwater movement remains down the topographic gradient and towards the eastern outlet of the valley at Picabo and the western outlet near Stanton Crossing.

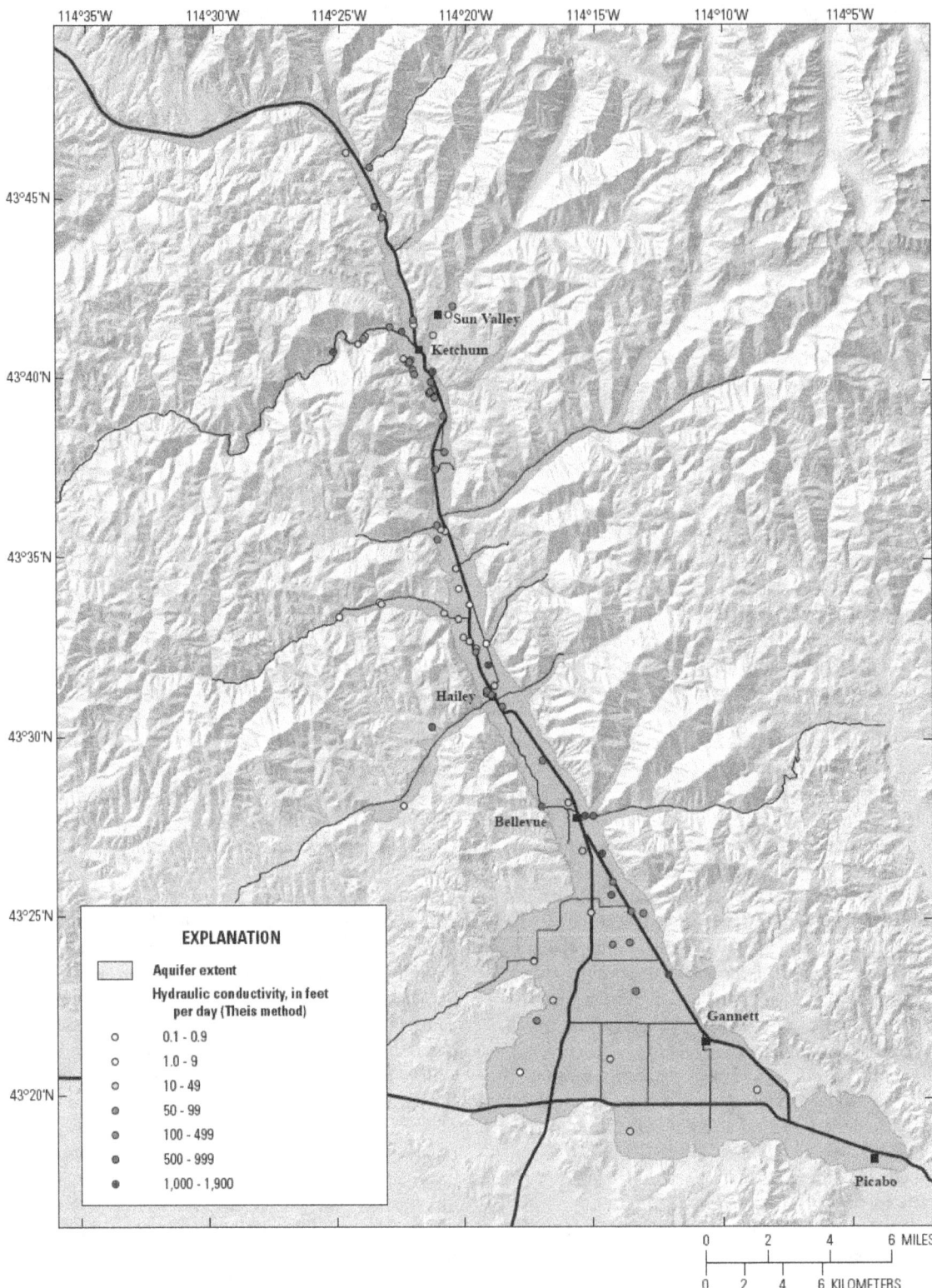

Figure 8.   Estimated hydraulic conductivity for selected wells completed in the Wood River Valley aquifer system and pre-Quaternary bedrock, Wood River Valley, south-central Idaho.

Table 2. Summary statistics, estimated hydraulic conductivity of 81 wells in the Wood River Valley aquifer system by the Thomasson and others (1960) and Theis and others (1963) methods.

[Includes one value from Smith (1959) and eight values from Frenzel (1989). **Abbreviation:** ft/d, foot per day]

	Aquifer condition	Number of wells	Hydraulic conductivity (ft/d)				
			High	Low	Mean	Median	Geometric mean
Thomasson and others (1960)	Unconfined	78	2,300	1.2	300	97	100
	Confined	3	84	34	52	37	47
Theis and others (1963)	Unconfined	78	1,900	0.33	250	71	69
	Confined	3	52	32	42	43	42

Table 3. Published ranges of hydraulic conductivity for unconsolidated sediment and basalt.

[**Abbreviation:** ft/d, foot per day]

Material	Range of hydraulic conductivity (ft/d)	Source
Clay	2.8×10^{-6}–0.13	Spitz and Moreno (1996, p. 346)
Silt, sandy silts, clayey sands, till	2.8×10^{-3}–0.28	Fetter (2001, p. 85)
Silty sands, fine sands	0.028–2.8	Fetter (2001, p. 85)
Well-sorted sands, glacial outwash	2.8–280	Fetter (2001, p. 85)
Sand	1.3–2.8×10^3	Spitz and Moreno (1996, p. 348)
Sand and gravel	27–650	Spitz and Moreno (1996, p. 348)
Gravel	130–2.8×10^5	Spitz and Moreno (1996, p. 347)
Basalt	5.3×10^{-6}–0.13	Spitz and Moreno (1996, p. 346)
Basalt, permeable	0.13–1.3×10^4	Spitz and Moreno (1996, p. 346)
Basaltic lava and sediments	510–5.1×10^4	Spitz and Moreno (1996, p. 346)
Basalt, thin, fractured, sediment interbeds	0.01–2.4×10^4	Ackerman and others (2006, p. 19)
Basalt, massive, fractured, sediment interbeds	6.5–1.4×10^4	Ackerman and others (2006, p. 19)

Hydrogeologic Controls on Groundwater/ Surface-Water Interaction

Using streamflow data from streamflow-gaging stations, Bartolino (2009) described the Big Wood River between the Big Wood River at Hailey station (13139500) and the Big Wood River at Stanton Crossing near Bellevue station (13140800) as a losing reach (fig. 1). The Big Wood River between the Big Wood River near Ketchum station (13135500) and the Hailey station, and Silver Creek were described as gaining reaches. Below Glendale Road, the reach from Hailey to Stanton Crossing typically goes dry during the summer months due to reduced natural flows and irrigation diversions. During periods when the bed of the Big Wood River goes dry, flow in the channel resumes above Stanton Crossing due to inflow from springs and seeps along the Big Wood River and its tributaries (including Willow Creek, which enters the Big Wood below the current (2012) Stanton Crossing station). Moreland (1977) attributed the presence of these springs and seeps to higher percentages of fine-grained sediment on the southern part of the Bellevue fan, causing a corresponding decrease in transmissivity that forces groundwater to the surface. Figure 8 and appendix E show only one estimated hydraulic conductivity value for the unconfined aquifer in this area: 2.2 ft/d (at well 801770, 01S 18E 10DD). As shown on 1:100,000 scale topographic maps, many springs and tributary heads roughly correspond to the 4,920 ft (1,500 m) contour line. Moreland (1977) applied the same explanation of fine-grained sediment to Silver Creek, which arises from a number of springs and seeps on the eastern side of the Bellevue fan. As with the Big Wood River, spring and tributary head altitudes generally occur at about 4,920 ft.

Subsurface Groundwater Flow Beneath Silver Creek and Stanton Crossing

Previous investigators have used a variety of methods, such as basin yield and the Darcy equation, to estimate subsurface outflow of water from the Wood River Valley aquifer system into the eastern Snake River Plain aquifer. In making such estimates, however, the earlier investigators have necessarily made a number of assumptions about the subsurface geology in these areas. The updated hydrogeologic framework described in the current report allows improved estimates of the magnitude of this subsurface flow.

The use of the Darcy equation to calculate groundwater flow requires values for hydraulic gradient and hydraulic conductivity that must then be multiplied by the cross-sectional area of the aquifer system in order to determine a cross-sectional flow rate. For the current analysis, the hydraulic gradient is taken as 0.00307 (between wells 17 and 20 of Skinner and others [2007], measured in October 2006). The distance across the surface of the Quaternary alluvium is taken as 12,700 ft, as measured along a north-south line (on

the 114° 06' 54" west meridian) between Idavada Volcanics to the north and Sun Valley Group rocks to the south. Both of these bedrock units are assumed to be virtually impermeable. On the basis of well-driller reports from the three wells at the Hayspur Fish Hatchery described in section, "Basalts and Associated Deposits," it is assumed that outflow takes place through a completely saturated, ellipse segment-shaped cross section filled by horizontal units of alluvium, basalt, and alluvium. Thicknesses of the upper alluvium, basalt, and lower alluvium were taken as 75 ft, 108 ft, and 76 ft, respectively (unit thicknesses were taken from the well-driller reports for 853072, 01S 19E 13DB and 715300 01S 19E 13BD). Hydraulic conductivity values of 69 ft/d for the two alluvial layers and 50 ft/d for the basalt are assumed (the geometric mean for the unconfined system in table 2 and the sole Theis and others (1963) hydraulic conductivity value for a basalt well in appendix E); the estimated Theis and others (1963) hydraulic conductivity value for well 853072 estimated for this report is 32 ft/d. These values lead to an estimated outflow of 4,000 acre-ft/yr, a value much lower than previous estimates: 38,000 acre-ft/yr (Smith, 1959), 11,800 acre-ft/yr (Brockway and Kahlown, 1994), 53,000 acre-ft/yr (Garabedian, 1992), 47,000 acre-ft/yr (Cosgrove and others, 2006), and 20,000 acre-ft/yr (Bartolino, 2009).

Groundwater outflow beneath Stanton Crossing to the Camas Prairie can be estimated using a similar approach. Aquifer cross sectional area was estimated along a 3,000 ft line drawn across alluvium between bedrock exposures and passing through well 826192 (01S 18E 22BB) and HVSR measurement point SpringCrk_SP08a with bedrock at 35 and 84 ft depth, respectively. A completely saturated, quadrilateral cross section composed of alluvium with a hydraulic conductivity of 69 ft/d is assumed. Because Skinner and others (2007) did not measure a groundwater level near Stanton Crossing in October 2006, a hydraulic gradient of 0.004249 was taken as the slope between the water-level measurement of 4,946 ft from well 7 (Skinner and others, 2007) and the Big Wood River at Stanton Crossing near Bellevue gage (13140800) stage of 4,832 ft on October 23, 2006 (K.D. Skinner, U.S. Geological Survey, unpub. data, 2011; U.S. Geological Survey, 2011). The resulting estimate is 300 acre-ft/yr. Smith (1959) estimated that groundwater outflow beneath Stanton Crossing was "relatively small" while Brockway and Grover (1978), Brockway and Kahlown (1994), and Wetzstein and others (1999) considered it "negligible." The outflow estimate presented here should be considered having some uncertainty because of the lack of groundwater-level measurements near Stanton Crossing. Furthermore, at high lake stages, backwater from Magic Reservoir may affect groundwater gradients through this area. The Magic Reservoir spillway elevation is 4,800 ft (the elevation shown on the 1986 Magic Reservoir East 7.5 minute quadrangle, converted to NAVD 88); currently (2012),the elevation of the water surface at the Big Wood River at Stanton Crossing near Bellevue gaging station (13140800) at 0.0 stage is 4,828 ft (K.D. Skinner, U.S. Geological Survey, unpub. data, 2011).

Bedrock Flow Systems and Thermal Springs

In the Wood River Valley and its tributaries, more than 250 wells are drilled into and (or) completed in the bedrock underlying the Quaternary sediment and basalt of the Wood River Valley aquifer system. It is unclear how much water these rocks contribute to these wells and whether they represent flow systems separate from the Wood River Valley aquifer system (like the hot springs of the area), downward percolation of water from the main aquifer system through fractures, wellbore storage of main aquifer system water in low yield areas, or some combination of the three. Well-driller reports commonly indicate competent, dry rock between saturated Quaternary alluvium and water-bearing units of older bedrock, suggesting that in some areas they represent separate aquifers.

Thermal springs in the Wood River Valley area are present in the tributary canyons on the west side of the Big Wood River in proximity to rocks of the Idaho batholith and Challis Volcanics. Although many hot springs exist, the largest are Russian John and Easley Hot Springs near the Big Wood River above the North Fork confluence, Guyer and Warfield Hot Springs in the Warm Springs Creek drainage, Clarendon Hot Springs in the Deer Creek drainage, and Hailey Hot Springs in the Croy Creek drainage (fig. 1). Flow to the springs is controlled by faults and fractures and, based on geochemical analyses, thermal water seems to have interacted with igneous rocks of the Idaho batholith and Challis Volcanics (Anderson and Bideganeta, 1985; Foley and Street, 1988; Street, 1990). Analysis of oxygen-18/deuterium isotopic ratios and carbon-14 data for thermal water from southwestern Idaho and the Idaho batholith led Young and Lewis (1982) to conclude that most of the thermal water had recharged during the Pleistocene (Young and Lewis, 1982). Young (1985) also mentioned the "possibility of recharge from recent precipitation at high altitudes." Oxygen-18/deuterium data for hot- and cold-water springs in the study were collected by Foley and Street (1988) and Street (1990), who, lacking any carbon-14 data, agreed that the thermal water probably was recharged during the Pleistocene or possibly more recently at high elevations.

The cities of Hailey and Bellevue derive much of their municipal water supply from springs in tributaries to the east of the Big Wood River—Hailey from the Indian Creek drainage and Bellevue from the Seaman's Creek drainage. Both springs issue from alluvium on the tributary valley floor in or adjacent to the respective streams. It is unclear whether these springs are fed by water from the alluvium, bedrock beneath the alluvium, or some combination of the two.

Data Needs and Suggestions for Further Study

The construction of a hydrogeologic framework requires characterization of the subsurface using data that are by nature sparse and incomplete. Although the collection of additional data is always desirable, time and funding constraints necessarily restrict such activities to the collection of what is considered to be the most useful information. The following types of data and information could be used to refine and improve the current understanding of the hydrogeologic framework of the Wood River Valley aquifer system:

- The consideration of water chemistry has evolved into a key requirement for the characterization of groundwater for several reasons. First and foremost is whether chemical constituents proscribe the intended use of the water, such as excess salinity limiting agricultural uses or elevated concentrations of naturally-occurring metals or radionuclides excluding human consumption. In the Wood River Valley, very few water samples have been analyzed for trace metals or radionuclides. Furthermore, groundwater in the Wood River Valley may be vulnerable to contamination by elevated concentrations of nutrients from wastewater disposal or agricultural practices. Sampling a number of wells for analysis of ambient water quality, especially nutrients, in the Wood River Valley aquifer system would assess potential anthropogenic impacts on groundwater quality as well as provide insight as to timing and location of groundwater recharge. Data on groundwater chemistry also can help describe the groundwater system itself—how water flows through the aquifer and the sources, amounts, and timing of recharge. This latter use of chemical data requires both standard chemical analyses as well as those for environmental tracers including radiocarbon, dissolved gasses, chlorofluorocarbons, and hydrogen isotopes.

- The hydraulic connection between the alluvial sediments and underlying bedrock remains poorly understood. Because these rocks are increasingly being utilized as a groundwater supply, it is important to better understand the source of this water, the degree to which these rocks are in hydraulic connection with the Wood River Valley aquifer system, long-term water availability, and the chemical suitability for proposed uses. Geochemical analysis of water from wells completed in these older rocks would build upon previous work for geothermal characterization.

- The most recent groundwater-level measurements representative of the Wood River Valley aquifer system as a whole were made in October 2006 and documented in Skinner and others (2007). A repeat of the measurements in the network established by Skinner and others (2007) would document any changes in water levels that might affect management of the aquifer system as well as provide data necessary for a groundwater-flow model. Measurement of water levels in additional wells (those not in the existing network) completed solely in bedrock underlying the aquifer system would help assess any hydraulic connection.

- Additional HVSR measurements on the Bellevue fan would better define the bedrock surface and incised paleostream channels that may act as conduits for subsurface flow from the aquifer system. Two areas on private property in particular lack data: the southwestern part of the fan and the area between Price and Punkin Center Roads.

- Groundwater/surface-water interaction of the Big Wood River between Glendale Road and Stanton Crossing needs to be understood and quantified. A series of streamflow measurements through this reach between late-winter and early-summer in conjunction with measurements of water levels in wells near the stream channel would both help quantify the amount of recharge in this reach and provide information on the hydraulic characteristics of the streambed and aquifer system. Access to the river through private property and permission to measure water levels will be necessary.

- The most accurate means of determining the hydraulic conductivity and other hydraulic properties of an aquifer is a multiple-well aquifer test. Such tests tend to be costly and difficult to perform for several reasons: observation wells are seldom located close enough to the pumped well to detect drawdown, thus dedicated observation wells need to be drilled; pumping at a constant rate for an extended period requires electricity and disposal of large volumes of water; and interference from other pumping wells complicates analysis. Despite these difficulties multiple-well aquifer tests enable more accurate determination of hydraulic properties of aquifers than do single-well tests.

Summary and Conclusions

The Wood River Valley contains most of the population of Blaine County and the cities of Sun Valley, Ketchum, Hailey, and Bellevue. This mountain valley is underlain by the primarily alluvial Wood River Valley aquifer system which consists of a single unconfined aquifer that underlies the entire valley, an underlying confined aquifer that is present only in the southernmost valley, and the confining unit that separates them. The entire population of the area depends on groundwater for domestic supply, either from domestic or municipal-supply wells, and rapid population growth since the 1970s has caused concern about the long-term sustainability of the groundwater resource.

As part of an ongoing USGS effort to characterize the groundwater resources of the Wood River Valley, this report describes the hydrogeologic framework of the Wood River Valley aquifer system. Descriptions of the geologic history, hydrogeologic framework, and further data needs are included in this document.

After the deposition or emplacement of the oldest rocks in or adjacent to the study area the subsequent series of geologic events is responsible for the distribution and hydrologic characteristics of water-bearing rocks in the Wood River Valley. The Paleozoic history in the area was characterized by the deposition of marine sedimentary rocks and orogeny. In Late Cretaceous time, the granitic Idaho batholith was intruded into older sedimentary rocks after which it was exhumed and then covered by the eruption of the Eocene Challis Volcanic Group. During Miocene time, the Idavada Volcanics were erupted in conjunction with the formation of the Snake River Plain and basin-and-range normal faulting.

During Quaternary time, the Wood River Valley aquifer system was created. In early-Pleistocene time, the ancestral Big Wood River flowed into a broad alluvial fan with its apex at Bellevue (the Bellevue fan) and exited the valley either to the east or west of the Timmerman and Picabo Hills. A series of lava flows then erupted which dammed and diverted the river between the eastern and western sides of the Timmerman and Picabo Hills, possibly multiple times, resulting in a sequence of fluvial and lacustrine sediments of some thickness beneath the Bellevue fan that interfinger with basalts to the southeast and possibly southwest. Two episodes of glaciation provided sediment ranging in size from clay to boulders. The Wood River Valley assumed its current form when the Big Wood River incised about 30 ft, probably from a combination of uplift and episodic climatic changes.

Although the majority of the Wood River Valley aquifer system is composed of Quaternary-age sediments and basalts of the Wood River Valley and its tributaries, the older igneous, sedimentary, or metamorphic rocks that underlie these Quaternary deposits also are used for water supply. It is unclear to what extent these rocks are hydraulically connected to the main part of Wood River Valley aquifer system and thus whether they are separate aquifers. Briefly, these pre-Quaternary rocks include:

- Paleozoic sedimentary rocks in and near the study area that provide water to wells and springs include the Phi Kappa and Trail Creek Formations (Ordovician and Silurian), the Milligen Formation (Devonian), and the Sun Valley Group including the Wood River Formation (Pennsylvanian-Permian) and the Dollarhide Formation (Permian). Lithologically these rocks range from limestones and shales to sandstones and conglomerate. Because these rocks are typically metamorphosed to some degree they have little primary porosity.

- Granitic intrusive rocks of the Late Cretaceous Idaho batholith are exposed on the western side of the Wood River Valley in a northwest-trending band of discontinuous outcrops and likely provide water to wells in these areas. These rocks have only secondary porosity due to fractures.

- Rocks of the Eocene Challis Volcanic Group are found in and adjacent to the study area and bear water to wells and springs. Hydraulic conductivity varies widely due to both primary and secondary porosity.

- Outcrops of the Miocene Idavada Volcanics are found in the southern part of the study area adjacent to the Bellevue fan. These rhyolitic tuffs and tuffaceous sediments have primary porosity and probably yield water to wells and springs in the area of the Timmerman and Picabo Hills.

Quaternary-age alluvial sediment and volcanic rocks are the primary sources of groundwater in the Wood River Valley aquifer system. These Quaternary deposits can be divided into three units: a coarse-grained sand and gravel unit, a fine-grained silt and clay unit, and a single basalt unit. The fine- and coarse-grained units were primarily deposited as alluvium derived from glaciation in the surrounding mountains and upper reaches of tributary canyons. The basalt unit is found in the southeastern Bellevue fan area and is composed of two flows of different ages.

A map of the approximate altitude of the pre-Quaternary bedrock and Quaternary basalt surface in the Wood River Valley generally mimics the land surface by decreasing down tributary canyons and the main valley from north to south. Altitudes range from more than 6,700 feet in Baker Creek to less than 4,600 feet in the central Bellevue fan. Quaternary sediment thickness ranges from less than a foot on main and tributary valley margins to about 350 feet in the central Bellevue fan. Estimated hydraulic conductivity in alluvium of the Wood River Valley, based on data from well-driller reports for 81 wells, ranges from 1,900 feet per day (ft/d) along Warm Springs Creek to less than 1 ft/d in upper Croy Canyon. A well completed in bedrock had an estimated hydraulic conductivity value of 10 ft/d, one well completed in basalt had a value of 50 ft/d, and three wells completed in the confined system had values ranging from 32 to 52 ft/d.

In general, the pattern of groundwater movement through the Wood River Valley aquifer system is relatively straightforward. Groundwater under unconfined conditions moves down valley to the Bellevue fan, where it either enters a deeper confined aquifer or remains in a shallow unconfined aquifer; the two aquifers appear to hydraulically reconnect in the area south of Gannett. Recharge is primarily from precipitation or streamflow in losing reaches and discharge is primarily through springs and seeps to streams or outflow from the aquifer system. Rerouting of surface water into a network of irrigation canals in the late 19th century, construction of groundwater wells, and increased demand have affected groundwater flow, but the overall direction of groundwater movement remains down the topographic gradient and towards the eastern outlet of the valley at Picabo and western outlet near Stanton Crossing.

Subsurface outflow of groundwater from the Wood River Valley aquifer system into the eastern Snake River Plain aquifer was estimated to be 4,000 acre-feet per year using data from three wells at the Hayspur Fish Hatchery. Groundwater outflow beneath Stanton Crossing to the Camas Prairie was estimated to be 300 acre-feet per year using data from a single well and a HVSR measurement.

Several data and information needs and potential studies were identified during this investigation that would provide for a more refined characterization of the hydrogeologic framework of the Wood River Valley aquifer system. These include:

- Groundwater-quality sampling and analysis would help characterize the suitability of water for intended uses, assess potential anthropogenic impacts, contribute to an understanding of how groundwater moves through the aquifer system, and help characterize groundwater flow in older bedrock.

- A repeat of groundwater-level measurements made in October 2006 would document any changes in water levels that might affect management of the aquifer system as well as provide data necessary for a groundwater-flow model. Measurement of water levels in additional wells (those not in the existing network) completed solely in bedrock underlying the aquifer system would help assess any hydraulic connection.

- The collection of additional horizontal-to-vertical spectral ratio (HSVR) data on the Bellevue fan would enable better definition of the bedrock surface and incised paleostream channels that may act as conduits for subsurface flow from the aquifer system.

- The understanding of groundwater/surface-water interaction of the Big Wood River between Glendale Road and Stanton Crossing could be enhanced and quantified through a series of streamflow measurements through this reach.

- Finally, multiple-well aquifer tests would allow aquifer properties to be determined more accurately than analysis of single-well tests.

Acknowledgments

The following individuals provided technical information, discussion, and suggestions during the preparation of this report: D.L. Garwood, R.M. Breckenridge, and R.S. Lewis of the Idaho Geological Survey, G.D. Thackray of Idaho State University, and J.W. Lane, E.B. Voytek, and D.L. Galloway of the USGS. Technical and field assistance was provided by K.D. Skinner, R.J. Weakland, A.B. Etheridge, and D.E. MacCoy of the USGS. Constructive technical reviews were provided by S.C. Kahle and J.C. Fisher of the USGS. Assistance and support was provided by the Blaine County Commissioners (as of January 2012): Tom Bowman, Angenie McCleary, and Lawrence Schoen. Funding and technical assistance was provided by the cooperators who helped fund the study and their representatives: Blaine County, City of Hailey, City of Ketchum, The Nature Conservancy, City of Sun Valley, Sun Valley Water and Sewer District, Blaine Soil Conservation District, and City of Bellevue. Finally, thanks to the citizens of the Wood River Valley who funded and supported this work.

References Cited

Ackerman, D.J., Rattray, G.W., Rousseau, J.P., Davis, L.C., and Orr, B.R., 2006, A conceptual model of ground-water flow in the eastern Snake River Plain aquifer at the Idaho National Laboratory and vicinity with implications for contaminant transport: U.S. Geological Survey Scientific Investigations Report 2006-5122, 62 p. (Also available at http://pubs.usgs.gov/sir/2006/5122/.)

Anderson, J.E., and Bideganeta, Kim, 1985, Geothermal investigations in Idaho, part 13, A preliminary geologic reconnaissance of the geothermal occurrences of the Wood River drainage area: Boise, Idaho Department of Water Resources Water Information Bulletin 30, Part 13, 49 p., accessed January 31, 2012, at http://www.idwr.idaho.gov/WaterInformation/Publications/wib/wib30p13-geothermal_wood_river_drainage_area.pdf.

Bartolino, J.R., 2009, Ground-water budgets for the Wood River Valley aquifer system, south-central Idaho: U.S. Geological Survey Scientific Investigations Report 2009-5016, 36 p. (Also available at http://pubs.usgs.gov/sir/2009/5016/.)

Beranek, L.P., Link, P.K., and Fanning, C.M., 2006, Miocene to Holocene landscape evolution of the western Snake River Plain region, Idaho: Using the SHRIMP detrital zircon provenance record to track eastward migration of the Yellowstone Hotspot: Geological Society of America Bulletin, v. 118, no. 9/10, p. 1027-1050.

Breckenridge, R.M., and Othberg, K.L., 2006, Surficial geologic map of the Wood River Valley area, Blaine County, Idaho: Moscow, Idaho Geological Survey Digital Web Map 54, scale 1:50,000, 1 sheet, accessed January 31, 2012, at http://www.idahogeology.org/PDF/Digital_Data_(D)/Digital_Web_Maps_(DWM)/Wood_River_Surf_DWM-54-M.pdf.

Breckenridge, R.M., Othberg, K.L., and Esser, R.P., 2003, New ages for Bellevue Formation—implications for Quaternary stratigraphy of the Wood River Valley, Idaho: Geological Society of America, Abstracts with Programs, v. 35, no. 6, p. 334, accessed January 31, 2012, at http://gsa.confex.com/gsa/2003AM/finalprogram/abstract_62331.htm.

Breckenridge, R.M., Sprenke, K.F., and Stryhas, B.A., 1984, List of Idaho earthquakes, 1972–1983: Moscow, Idaho Geological Survey Technical Report T-84-1, 41 p., accessed January 31, 2012, at http://www.idahogeology.org/PDF/Technical_Reports_(T)/TR-84-1.pdf.

Brockway, C.E., and Grover, K.P., 1978, Evaluation of urbanization and changes in land use on the water resources of mountain valleys: Moscow, University of Idaho Water Resources Research Institute, 104 p. plus appendix.

Brockway, C.E., and Kahlown, M.A., 1994, Hydrologic evaluation of the Big Wood River and Silver Creek watersheds Phase I: Kimberly, University of Idaho Water Resources Research Institute, Kimberly Research Center, 53 p. plus 5 appendices, accessed January 31, 2012, at http://conserveonline.org/docs/2004/06/hydrology_phase1_1994.pdf.

Brown, A.L., Jr., 2000, Summary Report—Hydrologic evaluation of the Big Wood River and Silver Creek watersheds: Ketchum, Hydronetics, 84 p.

Castellero, S., and Mulargia, F., 2009, V^{S30} estimates using constrained H/V measurements: Bulletin of the Seismological Society of America, v. 99, no. 2A, p. 761-773.

Castelin, P.M., and Chapman, S.L., 1972, Water resources of the Big Wood River-Silver Creek area, Blaine County, Idaho: Boise, Idaho Department of Water Administration, Water Information Bulletin 28, 44 p., accessed January 31, 2012, at http://www.idwr.idaho.gov/WaterInformation/Publications/wib/wib28-big_wood_river-silver_creek_area.pdf.

Castelin, P.M., and Winner, J.E., 1975, Effects of urbanization on the water resources of the Sun Valley-Ketchum area, Blaine County, Idaho: Boise, Idaho Department of Water Resources, Water Information Bulletin 40, 86 p., accessed January 31, 2012, at http://www.idwr.idaho.gov/WaterInformation/Publications/wib/wib40-sun_valley-ketchum_area.pdf.

Chapman, S.H., 1921, Water distribution and hydrometric work, Districts 7 and 11, Big and Little Wood Rivers: Shoshone, Annual Report Watermaster Districts 7 and 11, 21 p., 67 charts.

Cosgrove, D.M., Contor, B.A., and Johnson, G.S., 2006, Enhanced Snake Plain aquifer model—final report: Idaho Falls, University of Idaho Water Resources Research Institute, Technical Report 06-002, Eastern Snake Plain Aquifer Model Enhancement Project Scenario Document Number DDM-019, 120 p., plus tables, figures, plates, and appendices, accessed January 31, 2012, at http://www.if.uidaho.edu/%7ejohnson/FinalReport_ESPAM1_1.pdf.

DeGrey, Laura, Miller, Myles, and Link, P.K., 2011, Mesozoic Idaho batholith, Digital geology of Idaho, Digital atlas of Idaho, accessed January 31, 2012, at http://geology.isu.edu/Digital_Geology_Idaho/Module6/mod6.htm.

Di Giulio, G., Cornou, C., Ohrnberger, M., Wathelet, M., and Rovellii, A., 2006, Deriving wavefield characteristics and shear-velocity profiles from two-dimensional small-aperture arrays analysis of ambient vibrations in a small-size alluvial basin, Colfiorito, Italy: Bulletin of the Seismological Society of America, v. 96, no. 5, p. 1915-1933.

Dover, J.H., 1983, Geologic map and sections of the central Pioneer Mountains, Blaine and Custer Counties, central Idaho: U.S. Geological Survey Miscellaneous Investigations Series Map I-1319, scale 1:24,000, 2 sheets, accessed January 23, 2012, at http://ngmdb.usgs.gov/Prodesc/proddesc_9151 htm.

Eldridge, G.H., 1896, A geological reconnaissance across Idaho: U.S. Geological Survey Sixteenth Annual Report, Part 2, p. 264-271. (Also available at http://pubs.er.usgs.gov/pubs/ar/ar16_2.)

ESRI, 2001, ArcGIS Geostatistical Analyst—statistical tools for data exploration, modeling, and advanced surface generation: Redlands, Calif., ESRI White Paper J8647, 19 p., accessed January 31, 2012, at http://www.esri.com/library/whitepapers/pdfs/geostat.pdf.

Fetter, C.W., 2001, Applied hydrogeology, 4th ed.: Upper Saddle River, New Jersey, Prentice Hall, 598 p.

Foley, Duncan, and Street, L.V., 1988, Hydrothermal systems of the Wood River area, Idaho, in Link, P.K. and Hackett, W.R., eds., Guidebook to the geology of central and southern Idaho: Moscow, Idaho Geological Survey Bulletin 27, p. 109-126, accessed January 31, 2012, at http://www.idahogeology.org/PDF/Bulletins_(B)/B-27Full.pdf.

Forstall, Richard, ed., 1995, Idaho population of counties by decennial census: 1900 to 1990: Accessed January 31, 2012, at http://www.census.gov/population/cencounts/id190090.txt.

Frenzel, S.A., 1989, Water resources of the upper Big Wood River basin, Idaho: U.S. Geological Survey Water-Resources Investigation Report 89-4018, 47 p. (Also available at http://pubs.er.usgs.gov/pubs/wri/wri894018.)

Garabedian, S.P., 1992, Hydrology and digital simulation of the regional aquifer system, Eastern Snake River Plain, Idaho: U.S. Geological Survey Professional Paper 1408-F, 102 p., 10 plates in pocket. (Also available at http://pubs.er.usgs.gov/usgspubs/pp/pp1408F.)

Garwood, D.L., Kauffman, J.D., and Othberg, K.L., 2011, Geologic map of Picabo quadrangle, Blaine County, Idaho: Moscow, Idaho Geological Survey Digital Web Map 117, scale 1:24,000, 1 sheet, accessed January 31, 2012, at http://www.idahogeology.org/PDF/Digital_Data_(D)/Digital_Web_Maps_(DWM)/Picabo_DWM-131-m.pdf.

Garwood, D.L., Othberg, K.L., and Kauffman, J.D., 2010, Geologic map of Gannett quadrangle, Blaine County, Idaho: Moscow, Idaho Geological Survey Digital Web Map 117, scale 1:24,000, 1 sheet, accessed January 31, 2012, at http://www.idahogeology.org/PDF/Digital_Data_(D)/Digital_Web_Maps_(DWM)/Gannett_DWM-117-m.pdf.

Hackett, W.R., and Morgan, L.A., 1988, Explosive basaltic and rhyolitic volcanism of the eastern Snake River Plain, in Link, P.K., and Hackett, W.R., eds., Guidebook to the geology of central and southern Idaho: Moscow, Idaho Geological Survey Bulletin 27, p. 283-301, accessed January 31, 2012, at http://www.idahogeology.org/PDF/Bulletins_(B)/B-27Full.pdf.

Hall, W.E., 1985, Stratigraphy and mineralization in middle and upper Paleozoic rocks of the central Idaho black shale mineral belt, *in* McIntyre, D.H., ed., Symposium on the geology and mineral deposits of the Challis 1 degree by 2 degree quadrangle, Idaho: U.S. Geological Survey Bulletin 1658, chapter J, p. 117-131. (Also available at http://pubs.usgs.gov/bul/1658a-s/report.pdf.)

Hall, W.E., Batchelder, J.N., and Tschanz, C.N., 1978, Preliminary geologic map of the Sun Valley 7.5-minute Quadrangle, Idaho: U.S. Geological Survey Open-File Report 78-1058, scale 1:24,000, 1 sheet. (Also available at http://pubs.usgs.gov/of/1978/1058/plate-1.pdf.)

Hughes, S.S., Smith, R.P., Hackett, W.R., and Anderson, S.R., 1999, Mafic volcanism and environmental geology of the eastern Snake River Plain, Idaho, *in* Hughes, S.S. and Thackray, G.D., eds., Guidebook to the geology of eastern Idaho: Idaho Museum of Natural History, p. 143-168, accessed January 31, 2012, at http://geology.isu.edu/ Digital_Geology_Idaho/Module11/Hughesetal1999.pdf.

Ibs-von Seht, M., and Wohlenberg, J., 1999, Microtremor measurements used to map thickness of soft sediments: Bulletin of the Seismological Society of America, v. 89, no. 1, p. 250-259.

Idaho Department of Water Resources, 2012, Well driller reports (logs): Idaho Department of Water Resources database, accessed January 31, 2012, at http://www. idwr.idaho.gov/WaterManagement/WellInformation/ DrillerReports/dr_default.htm.

Idaho Geological Survey, 2012, Geologic hazards— earthquakes—Historical earthquakes in Idaho, 1880–1989: Idaho Geological Survey, accessed January 31, 2012, at http://www.idahogeology.org/Services/GeologicHazards/ Earthquakes/historiceqs.html.

Ingram, R.L., 1989, Grain-size scale used by American geologists, AGI data sheet 29.1, *in*, Dutro, J.T., Jr., Dietrich, R.V., and Foose, R.M., eds., AGI data sheets, 3rd ed.: Alexandria, Va., American Geological Institute, variously paged.

Johnson, K.M., Lewis, R.S., Bennett, E.H., and Kiilsgaard, T.H., 1988, Cretaceous and Tertiary intrusive rocks of south-central Idaho, *in* Link, P.K., and Hackett, W.R., eds., Guidebook to the geology of central and southern Idaho: Moscow, Idaho Geological Survey Bulletin 27, p. 55-86, accessed January 31, 2012, at http://www.idahogeology.org/ PDF/Bulletins_(B)/B-27Full.pdf.

Jones, R.P., 1952, Evaluation of streamflow records in the Big Wood River basin, Blaine County, Idaho: U.S. Geological Survey Circular 192, 59 p., 1 plate in pocket. (Also available at http://pubs.er.usgs.gov/usgspubs/cir/cir192.)

Kauffman, J.D., and Othberg, K.L., 2007, Geologic map of the Magic Reservoir east quadrangle, Blaine and Camas Counties, Idaho: Moscow, Idaho Geological Survey Digital Web Map 82, scale 1:50,000, 1 sheet, accessed January 31, 2012, at http://www.idahogeology.org/PDF/Digital_Data_ (D)/Digital_Web_Maps_(DWM)/magic_res_east_dwm- 82-m.pdf.

Kiilsgaard, T.H., Lewis, R.S., and Bennett, E.H., 2001, Plutonic and hypabyssal rocks of the Hailey 1 degree x 2 degrees quadrangle, Idaho: U.S. Geological Survey Bulletin 2064-U, 18 p. (Also available at http://pubs.usgs.gov/bul/ b2064-u/b2064-u.pdf.)

Lane, J.W., Jr., White, E.A., Steele, G.V., and Cannia, J.C., 2008, Estimation of bedrock depth using the horizontal-to-vertical (H/V) ambient-noise seismic method, *in* Symposium on the application of geophysics to engineering and environmental problems, April 6-10, 2008, Philadelphia, Pennsylvania, proceedings: Denver, Environmental and Engineering Geophysical Society, 13 p.

Lewis, R.S., 1989, Plutonic rocks in the southeastern part of the Idaho batholith and their relationship to mineralization, *in* Winkler, G.R., Soulliere, S.J., Worl, R.G., and Johnson, K.M., eds., Geology and mineral deposits of the Hailey and western Idaho Falls 1 degree x 2 degree quadrangles, Idaho: U.S. Geological Survey Open-File Report 89-639, p. 33-34. (Also available at http://pubs.usgs.gov/of/1989/0639/report. pdf.)

Lindgren, Waldemar, 1900, The gold and silver veins of Silver City, De Lemar, and other mining districts in Idaho: U.S. Geological Survey Twentieth Annual Report, Part 3, p. 207-209.

Link, P.K., 2011, Neogene Snake River Plain-Yellowstone Volcanic Province, Digital geology of Idaho, Digital atlas of Idaho, accessed January 31, 2012, at http://geology.isu.edu/ Digital_Geology_Idaho/Module11/mod11.htm.

Link, P.K., and Rodgers, D.W., compilers, 1995, Geologic map of the northeastern part of the Hailey 1 degree x 2 degree quadrangle, south-central Idaho, plate, *in* Worl, R.G., Link, P.K., Winkler, G.R., and Johnson, K.M., eds., Geology and mineral resources of the Hailey 1 degree × 2 degree quadrangle and the western part of the Idaho Falls 1 degree × 2 degree quadrangle, Idaho: U.S. Geological Survey Bulletin 2064-B, plate 1. (Also available at http:// pubs.usgs.gov/bul/2064a-r/report.pdf.)

Link, P.K., Fanning, C.M., and Beranek, L.P., 2005, Reliability and longitudinal change of detrital-zircon age spectra in the Snake River system, Idaho and Wyoming—An example of reproducing the bumpy barcode: Sedimentary Geology, v. 182, p. 101–142.

Link, P.K., Mahoney, J.B., Bruner, D.J., Batatian, L.D., Wilson, Eric, and Williams, F.J.C., 1995a, Stratigraphic setting of sediment-hosted mineral deposits in the eastern part of the Hailey 1 degree x 2 degree quadrangle and part of the southern part of the Challis 1 degree x 2 degrees quadrangle, south-central Idaho, *in* Worl, R.G., Link, P.K., Winkler, G.R., and Johnson, K.M., eds., Geology and mineral resources of the Hailey 1 degree × 2 degree quadrangle and the western part of the Idaho Falls 1 degree × 2 degree quadrangle, Idaho: U.S. Geological Survey Bulletin 2064-A, p. C1-C33. (Also available at http://pubs.usgs.gov/bul/2064a-r/report.pdf.)

Link, P.K., Mahoney, J.B., Bruner, D.J., Batatian, L.D., Wilson, Eric, and Williams, F.J.C., 1995b, Geologic map of outcrop areas of sedimentary units in the Eastern part of the Hailey 1 degree x 2 degrees quadrangle, South-Central Idaho: U.S. Geological Survey Bulletin 2064-C, 1 plate, scale 1:250,000. (Also available at http://pubs.usgs.gov/bul/b2064-c/.)

Link, P.K., Skipp, Betty, Hait, M.H., Jr., Janecke, Susanne, and Burton, B.R., 1988, Structural and stratigraphic transect of south-central Idaho—A field guide to the Lost River, White Knob, Pioneer, Boulder, and Smoky Mountains, *in* Link, P.K., and Hackett, W.R., eds., Guidebook to the geology of central and southern Idaho: Moscow, Idaho Geological Survey Bulletin 27, p. 5-42, accessed January 31, 2012, at http://www.idahogeology.org/PDF/Bulletins_(B)/B-27Full.pdf.

Luttrell, S.P., and Brockway, C.E., 1984, Impacts of individual on-site sewage disposal facilities on mountain valleys—Phase II—Water-quality considerations: Moscow, Idaho Water and Energy Resources Research Institute, University of Idaho, Research Technical Completion Report WRIP/371403, 74 p.

Mace, R.E., 2001, Estimating transmissivity using specific-capacity data: Austin, Tex. Bureau of Economic Geology Geological Circular 01-02, 44 p.

Mahoney J.B., Link, P.K., Burton, B.R., Geslin, J.K., and O'Brien, J.P., 1991, Pennsylvanian and Permian Sun Valley Group, Wood River Basin, south-central Idaho, *in* Cooper, J.D., and Stevens, C.H., eds., Paleozoic paleogeography of the western United States-II: Los Angeles, Pacific Section, Society of Economic Paleontologists and Mineralogists (Society for Sedimentary Geology), Publication 67, v. 2, p. 551-579

Malde, H.E., and Powers, H.A., 1962, Upper Cenozoic stratigraphy of western Snake River Plain, Idaho: Geological Society of America Bulletin, v. 73, no. 10, p. 1197-1220.

Micromed S.p.A., 2011, Grilla software: Micromet S.p.A., accessed January 31, 2012, at http://www.tromino.eu.

Moreland, J.A., 1977, Ground water-surface water relations in the Silver Creek area, Blaine County, Idaho: Boise, Idaho Department of Water Resources, Water Information Bulletin 44, 42 p., 5 plates in pocket, accessed January 31, 2012, at http://www.idwr.idaho.gov/WaterInformation/Publications/wib/wib44-weiser_river_basin.pdf. Also published as U.S. Geological Survey Open-File report 77-456, 66 p., plates in pocket. (Also available http://pubs.er.usgs.gov/pubs/ofr/ofr77456.)

Moye, F.J., Hackett, W.R., Blakley, J.D., and Snider, L.G., 1988, Regional geologic setting and volcanic stratigraphy of the Challis volcanic field, central Idaho, *in* Link, P.K., and Hackett, W.R., eds., Guidebook to the geology of central and southern Idaho: Moscow, Idaho Geological Survey Bulletin 27, p. 87-97, accessed January 31, 2012, at http://www.idahogeology.org/PDF/Bulletins_(B)/B-27Full.pdf.

Nakamura, Y., 1989, A method for dynamics characteristics estimations of subsurface using microtremors on the ground surface: Quarterly Report, RTRI, Japan, v. 30, p. 25-33.

National Climatic Data Center, 2011, Climate monitoring: National Oceanic and Atmospheric Administration database, accessed January 31, 2012, at http://www.ncdc.noaa.gov/climate-monitoring/.

Neuendorf, K.K.E., Mehl, J.P., Jr., and Jackson, J.A., eds., 2011, Glossary of Geology, Fifth Edition (revised): Alexandria, Va., American Geological Institute, 800 p.

Parolai, S., Bormann, P., and Milkereit, C., 2002, New relationships between Vs, thickness of sediments, and resonance frequency calculated by the H/V ratio of seismic noise for the Cologne area (Germany): Bulletin of the Seismological Society of America, v. 92, no. 6, p. 2521-2527.

Pearce, Suzanne, Schlieder, Gunnar, Evenson, E.B., 1988, Field guides to the Quaternary geology of central Idaho—Part A. Glacial deposits of the Big Wood River Valley, *in* Link, P.K., and Hackett, W.R., eds., Guidebook to the geology of central and southern Idaho: Moscow, Idaho Geological Survey Bulletin 27, p. 203-207, accessed January 31, 2012, at http://www.idahogeology.org/PDF/Bulletins_(B)/B-27Full.pdf.

Rodgers, D.W., Link, P.K., and Huerta, A.D., 1995, Structural framework of mineral deposits hosted by Paleozoic rocks in the northeastern part of the Hailey 1 degree × 2 degree quadrangle, south-central Idaho, *in* Worl, R.G., Link, P.K., Winkler, G.R., and Johnson, K.M., eds., Geology and mineral resources of the Hailey 1 degree × 2 degree quadrangle and the western part of the Idaho Falls 1 degree × 2 degree quadrangle, Idaho: U.S. Geological Survey Bulletin 2064-A, p. B1-B18. (Also available at http://pubs.usgs.gov/bul/2064a-r/report.pdf.)

Sanford, R.F., 2005, Geology and stratigraphy of the Challis Volcanic Group and related rocks, Little Wood River area, south-central Idaho, *with a section on* Geochronology by Lawrence W. Snee: U.S. Geological Survey Bulletin 2064-II, 22 p. (Also available at http://pubs.usgs.gov/bul/2064/ii/pdf/B2064-II.pdf.)

Schmidt, D.L., 1962, Quaternary geology of the Bellevue area in Blaine and Camas Counties, Idaho: U.S. Geological Survey Open-File Report 62-120, 127 p. (Also available at http://pubs.er.usgs.gov/pubs/ofr/ofr62120.)

Skinner, K.D., Bartolino, J.R., and Tranmer, A.W., 2007, Water-resource trends and comparisons between partial development and October 2006 hydrologic conditions, Wood River Valley, south-central Idaho: U.S. Geological Survey Scientific Investigations Report 2007-5258, 31 p., 4 pls., 1 app. (Also available at http://pubs.er.usgs.gov/usgspubs/sir/sir20075258.)

Smith, R.O., 1959, Ground-water resources of the middle Big Wood River-Silver Creek area, Blaine County, Idaho: U.S. Geological Survey Water-Supply Paper 1478, 61 p., 2 plates in pocket. (Also available at http://pubs.er.usgs.gov/usgspubs/wsp/wsp1478.)

Smith, R.O., 1960, Geohydrologic evaluation of streamflow records in the Big Wood River basin, Idaho: U.S. Geological Survey Water-Supply Paper 1479, 64 p., 4 plates in pocket. (Also available at http://pubs.er.usgs.gov/pubs/wsp/wsp1479.)

Spence, C.C., 1999, For Wood River or bust—Idaho's silver boom of the 1880s: Moscow, University of Idaho Press, 260 p.

Spitz, Karlheinz, and Moreno, Joanna, 1996, A practical guide to groundwater and solute transport modeling: New York, John Wiley and Sons, 461 p.

Stearns, H.T., Crandall, Lynn, and Steward, W.G., 1938, Geology and ground-water resources of the Snake River Plain in southeastern Idaho: U.S. Geological Survey Water-Supply Paper 774, 268 p., 6 plates in pocket. (Also available at http://pubs.er.usgs.gov/pubs/wsp/wsp774.)

Street, L.V., 1990, Geothermal investigations in Idaho, part 17, Geothermal resource analysis in the Big Wood River valley, Blaine County, Idaho: Boise, Idaho Department of Water Resources, Water Information Bulletin 30, part 17, 26 p., 1 plate in pocket, accessed January 31, 2012, at http://www.idwr.idaho.gov/WaterInformation/Publications/wib/wib30p17-geothermal_big_wood_river_valley.pdf.

Theis, C.V., Brown, R.H., and Meyer, R.R., 1963, Estimating the transmissibility of aquifers from the specific capacity of wells, *in* Bentall, Ray, ed., Methods of determining permeability, transmissibility and drawdown: U.S. Geological Survey Water-Supply Paper 1536-I, p. 331-341. (Also available at http://pubs.usgs.gov/wsp/1536i/report.pdf.)

Thomasson, H.G. Jr., Olmstead, F.H., and LeRoux, E.F., 1960, Geology, water resources and usable ground-water storage capacity of part of Solano County, California: U.S. Geological Survey Water-Supply Paper 1464, 693 p. 19 plates in pocket. (Also available at http://pubs.usgs.gov/wsp/1464/report.pdf.)

Turner, R.J.W., and Otto, B.R., 1995, Structural and stratigraphic setting of the Triumph stratiform zinc-lead-silver deposit, Devonian Milligen Formation, central Idaho, *in* Worl, R.G., Link, P.K., Winkler, G.R., and Johnson, K.M., eds., Geology and mineral resources of the Hailey 1 degree × 2 degree quadrangle and the western part of the Idaho Falls 1 degree × 2 degree quadrangle, Idaho: U.S. Geological Survey Bulletin 2064-A, p. C1-C33. (Also available at http://pubs.usgs.gov/bul/2064a-r/report.pdf.)

Umpleby, J.B., Westgate, L.G., Ross, C.P., and Hewett, D.F., 1930, Geology and ore deposits of the Wood River region, Idaho, with a description of the Minnie Moore and near-by mines: U.S. Geological Survey Bulletin 814, 250 p. (Also available at http://pubs.usgs.gov/bul/0814/report.pdf.)

U.S. Census Bureau, 2011, State and county Quickfacts, Blaine County, Idaho: U.S. Census Bureau database, accessed August 5, 2011, at http://quickfacts.census.gov/qfd/states/16/16013.html.

U.S. Geological Survey, 2011, USGS Surface-Water Data for Idaho: U.S. Geological Survey database, accessed January 31, 2012, at http://waterdata.usgs.gov/id/nwis/sw.

U.S. Geological Survey Geologic Names Committee, 2010, Divisions of geologic time-major chronostratigraphic and geochronologic units: U.S. Geological Survey Fact Sheet 2010–3059, 2 p. (Also available at http://pubs.usgs.gov/fs/2010/3059/pdf/FS10-3059.pdf.)

Vallier, T.L., and Brooks, H.C., 1987, The Idaho batholith and its border zone—A regional perspective, *in* Vallier, T.L. and Brooks, H.C., eds., Geology of the Blue Mountains Region of Oregon, Idaho, and Washington—The Idaho batholith and its border zone: U.S. Geological Survey Professional Paper 1436, p. 1–8. (Also available at http://pubs.usgs.gov/pp/1438/report.pdf.)

Wathelet, M., 2011, Geopsy project: Accessed January 31, 2012, at http://www.geopsy.org/.

Wathelet, M., Jongmans, D., Ohrnberger, M., and Bonnefoy-Claudet, S., 2008, Array performances for ambient vibrations on a shallow structure and consequences over *Vs* inversion: Journal of Seismology, v. 12, no. , p. 1–19.

Wetzstein, A.B., Robinson, C.W., and Brockway, C.E., 1999, Hydrologic evaluation of the Big Wood River and Silver Creek watersheds, phase II: Kimberly, University of Idaho Water Resources Research Institute, Kimberly Research Center, 136 p., accessed January 31, 2012, at http://conserveonline.org/docs/2004/06/hydrology_phase2_1999.pdf.

Whitehead, R.L., 1992, Geohydrologic framework of the Snake River Plain regional aquifer system, Idaho and eastern Oregon: U.S. Geological Survey Professional Paper 1408-B, 32 p., 6 plates in pocket. (Also available at http://pubs.usgs.gov/pp/1408b/report.pdf.)

Worl, R.G., and Johnson, K.M., 1995a, Geology and mineral deposits of the Hailey 1 degree x 2 degree quadrangle and the western part of the Idaho Falls 1 degree × 2 degree quadrangle, south-central Idaho—An overview, *in* Worl, R.G., Link, P.K., Winkler, G.R., and Johnson, K.M., eds., Geology and mineral resources of the Hailey 1 degree × 2 degree quadrangle and the western part of the Idaho Falls 1 degree × 2 degree quadrangle, Idaho: U.S. Geological Survey Bulletin 2064-A, p. A1-A21. (Also available at http://pubs.usgs.gov/bul/2064a-r/report.pdf.)

Worl, R.G., and Johnson, K.M., 1995b, Map showing geologic terranes of the Hailey 1 degree x 2 degrees quadrangle and the western part of the Idaho Falls 1 degree × 2 degrees quadrangle, south-central Idaho: U.S. Geological Survey Bulletin 2064-A, 1 plate, scale 1:250,000. (Also available at http://pubs.usgs.gov/bul/b2064-a/.)

Worl, R.G., Kiilsgaard, T.H., Bennett, E.H., Link, P.K., Lewis, R.S., Mitchell, V.E., Johnson, K.M., and Snyder, L.D., 1991, Geologic map of the Hailey 1 degree × 2 degree Quadrangle, Idaho: U.S. Geological Survey Open-File Report 91-340, 1 sheet. (Also available at http://pubs.er.usgs.gov/pubs/ofr/ofr91340.) Also published as Worl, R.G., Kiilsgaard, T.H., Bennett, E.H., Link, P.K., Lewis, R.S., Mitchell, V.E., Johnson, K.M., and Snyder, L.D., 1991, Geologic map of the Hailey 1 degree × 2 degree Quadrangle, Idaho: Moscow, Idaho Geological Survey, Geologic Map GM-10, scale 1:250,000, 1 sheet., accessed January 31, 2012, at http://www.idahogeology.org/PDF/Geological_Maps_(GM)/GM-10.pdf.

Wust, S.L., and Link, P.K., 1988, Field guide to the Pioneer Mountains core complex, south-central Idaho, *in* Link, P.K., and Hackett, W.R., eds., Guidebook to the geology of central and southern Idaho: Moscow, Idaho Geological Survey Bulletin 27, p. 43–54, accessed January 31, 2012, at http://www.idahogeology.org/PDF/Bulletins_(B)/B-27Full.pdf.

Young, H.W., 1985, Geochemistry and hydrology of thermal springs in the Idaho batholith and adjacent areas, central Idaho: U.S. Geological Survey Water-Resources Investigations Report 85-4172, 44 p. (Also available at http://pubs.usgs.gov/wri/1985/4172/report.pdf.)

Young, H.W., and Lewis, R.E., 1982, Hydrology and geochemistry of thermal ground water in southwestern Idaho and north-central Nevada: U.S. Geological Survey Professional Paper 1044-J, 20 p., 2 plates in pocket. (Also available at http://pubs.usgs.gov/pp/1044j/report.pdf.)

Glossary

The reader is referred to Neuendorf and others (2011) for more complete and technical definitions.

Alluvial fan   An open fan-shaped or cone-shaped mass of sediment deposited by streams at canyon mouths along a mountain front.

Alluvium   A general term for sediment deposited by a stream or other running water; typically, a late Cenozoic age is implied.

Andesite   A typically gray to black extrusive igneous rock with the same general chemical composition as its intrusive equivalent, diorite. Andesitic magmas have intermediate viscosity and eruptions can be explosive (such as at Krakatoa).

Aquifer   A geologic formation, group of formations, or part of a formation that contains a sufficient amount of saturated permeable material (for example, soil, sand, gravel and (or) rock) to yield significant quantities of water to wells and springs.

Aquifer system   An aquifer system is two or more aquifers that are separated (at least locally) by impermeable rock or sediment units but function together as an aquifer with regional or sub-regional extent.

Aquifer test   A procedure to estimate the hydraulic properties of an aquifer by removing (or adding) known quantities of water from a well and observing the resulting changes in water levels in the pumped well (single-well aquifer test) and sometimes additional non-pumping wells (multiple-well aquifer test). The most commonly determined hydraulic parameters are transmissivity, hydraulic conductivity, and specific storage.

Argillite   A weakly-metamorphosed claystone, siltstone, or shale.

Basalt   A typically black extrusive igneous rock with the same general chemical composition as its intrusive equivalent, gabbro. Basaltic magmas generally have a low viscosity and thus eruptions tend to be relatively gentle, or non-explosive (such as Kilauea).

Basin-and-range   A regional topography composed of a series of faulted blocks in which the uplifted blocks form mountain ranges and the intervening downthrown blocks form basins filled with sediment eroded from the adjoining mountain blocks. Also refers to the physiographic province of the western United States dominated by this topography.

Batholith   An irregularly shaped body of intrusive igneous rock with an exposed surface greater than about 40 square miles.

Bedrock   The consolidated rock underlying soil or unconsolidated sediment.

Biotite   A soft, platy silicate mineral; a dark mica formed in igneous or metamorphic rock. It is commonly found in sedimentary rock although it easily weathers to clay.

Breccia   A cemented sedimentary or volcanic rock composed mostly of angular rock fragments in a finer-grained matrix.

Carbonaceous   Sedimentary rock containing organic matter.

Chert   A sedimentary rock composed of micrometer-scale silica crystals; usually it is synonymous with flint.

Clay   The finest class of detrital particles with a diameter less than 0.00016 inch; it is formed from the weathering of silicate minerals. Grade names are very fine, fine, medium, and coarse. A mixture of silt and clay is mud.

Colluvial transport   The process by which poorly sorted soil, sediment, or rock fragments are transported by rainwash, sheetwash, or downslope creep.

Confined aquifer   An aquifer bounded above and below by confining units and completely filled with water under pressure. Synonymous with artesian aquifer.

Confining unit   An impermeable or distinctly less permeable geologic unit within an aquifer or bounding one or more aquifers. Synonymous with confining bed.

Conglomerate   A cemented sedimentary rock composed mostly of gravel-size rounded rock fragments in a finer-grained matrix.

Dacite   A typically light gray extrusive igneous rock with the same general chemical composition as its intrusive equivalent, granodiorite. Dacitic magmas have high viscosity and eruptions tend to be explosive or highly explosive (such as Mount St. Helens).

Darcy equation   An equation defining the volume of water flowing through a unit of surface area of the aquifer (area); units simplify to area per time but are often expressed in other forms. It is the fundamental equation used to describe groundwater flow. Also known as Darcy's law.

Diamictite   An unsorted or poorly-sorted sedimentary or volcanic rock with particles of many sizes. Examples are poorly-sorted conglomerates or breccias; they may form in a variety of environments such as glaciers, mudslides or landslides, and submarine landslides.

Digital elevation model (DEM)   Digital data representing a three- dimensional surface, usually the land surface. It constitutes a gridded series of points on the surface describing location and altitude. The spacing or resolution of the grid is specified as a length; common resolutions in the United States are 5-, 10-, or 30-meters.

Dike   A planar igneous intrusion that cuts across the bedding or grain of the intruded rock.

Diorite   A dark intrusive igneous rock composed primarily of the silicate minerals quartz, plagioclase feldspar, and alkali feldspar as well as abundant dark silicate minerals such as biotite and hornblende. Of the total volume of quartz, plagioclase feldspar, and alkali feldspar in the rock, granodiorite contains 0–5 percent quartz, and more than 90 percent of the feldspar is plagioclase feldspar. The plagioclase contains more sodium than calcium. The equivalent extrusive rock is andesite.

Dip   The maximum angle that a geologic surface (such as a fault plane or bedding) makes with the horizontal; it is measured perpendicularly to the strike.

Dolomite   A mineral composed mostly of calcium magnesium carbonate; a rock composed mostly of dolomite is also a dolomite (or dolostone). Dolostones and limestones usually cannot be distinguished visually.

Ephemeral stream   A stream that flows only occasionally, usually in direct response to precipitation.

Extrusive rock   A type of igneous rock formed as molten material cools on the surface. Because extrusive rock cools relatively quickly it tends to have smaller mineral crystals than intrusive (irruptive) rock. Synonymous with volcanic rock.

Fault   Faults mark parts of the Earth's crust that have broken and where the two sides have slid past each other across the break; this relative motion can be vertical (dip-slip), horizontal (strike-slip), or a combination of both.

Fluvial deposit   A sedimentary deposit transported by, suspended within, or deposited by a stream or river.

Fold   A deformation of rock or other geologic structure that has been bent or otherwise warped by geologic activity.

Gabbro   A black intrusive igneous rock primarily composed of the silicate minerals quartz, plagioclase feldspar, and alkali feldspar as well as abundant dark silicate minerals such as pyroxene and olivine. Of the total volume of quartz, plagioclase feldspar, and alkali feldspar in the rock, gabbro contains 0–5 percent quartz, and more than 90 percent of the feldspar is plagioclase feldspar. The plagioclase contains more calcium than sodium. The equivalent extrusive rock is basalt.

Gaining stream   The interaction between a stream and an aquifer can usually be described by whether the stream gains water from or loses water to the aquifer. A gaining stream receives or gains flow from groundwater. A losing stream is one that loses or contributes flow to groundwater.

Geologic section   A vertical profile of subsurface geology based on geologic mapping of outcrops, borehole logs, geophysical data, and manmade excavations such as quarries or roadcuts.

Geologic time   Scientists currently understand the Earth to be about 4.7 billion years old. To facilitate the study of rocks and their features, geologists have divided this geologic time into a hierarchical system of units characterized by distinct assemblages of rock types and fossils (see table 1 in this report).

Geostatistics   The application of statistical methods to geology, primarily to assess properties that may be physically continuous. It includes such techniques as kriging and cokriging.

Geographic information system (GIS):   A computer program and corresponding databases that allow for the integration and analysis of spatial data.

Glacial outwash   Sediment deposited by streams of meltwater flowing from a glacier.

Glaciation   The formation, movement, and recession of glaciers or ice sheets resulting in both erosion and deposition.

Granite   A light-colored intrusive igneous rock composed of the silicate minerals quartz, plagioclase feldspar, and alkali feldspar as well as minor amounts of dark silicate minerals such as mica and hornblende. Of the total volume of quartz, plagioclase feldspar, and alkali feldspar in the rock, granite contains 20-60 percent quartz, and 10–65 percent of the feldspar is plagioclase feldspar. The equivalent extrusive rock is rhyolite.

Granodiorite   A grayish intrusive igneous rock composed of the silicate minerals quartz, plagioclase feldspar, and alkali feldspar as well as abundant dark silicate minerals such as biotite and hornblende. Of the total volume of quartz, plagioclase feldspar, and alkali feldspar in the rock, granodiorite contains 20–60 percent quartz, and 65–90 percent of the feldspar is plagioclase feldspar. The equivalent extrusive rock is dacite.

Gravel   The coarsest class of rock fragments or detrital particles with a diameter larger than 0.08 inch. It includes pebbles, cobbles, and boulders.

Groundwater-flow model   In general, a model is a simplified representation of the appearance or operation of a real object or system; groundwater-flow models attempt to reproduce, or simulate, the operation of a real groundwater system with a mathematical counterpart (a mathematical model). Mathematical models may use different methods to simulate groundwater-flow systems.

Hanging valley   An inactive tributary valley whose mouth is relatively higher than the floor of a larger steep-sided valley; commonly, but not always, an indicator of glaciation.

Hornblende   A hard, dark, prismatic silicate mineral. It is usually formed in metamorphic rock although it can be an important constituent of igneous rocks such as basalt.

Hot spot   An area of volcanic activity, typically 60 to120 miles across, that persists for tens of millions of years; it is the surface expression of a mantle plume. While hot spots appear to migrate over time, in actuality they remain stationary as tectonic plates migrate over them. A present-day example is the Yellowstone area in western Wyoming, USA.

Hydraulic conductivity   A measure of the rate at which a porous medium transmits water.

Hydraulic head   In groundwater, the sum of the elevation, pressure, and velocity heads (total head) where the velocity head is assumed to be zero. Hydraulic head is represented by the potentiometric surface.

Hydrogeologic framework   The description and mapping of thickness, extent, hydraulic properties, and boundaries of the hydrogeologic units that constitute an aquifer.

Hydrogeologic unit   A distinct soil or rock unit or zone which by virtue of its hydraulic properties has a distinct influence on the storage or movement of groundwater. Synonymous with hydrostratigraphic unit.

Igneous rock   Rock formed by the cooling and crystallization of molten material, such as magma.

Infiltration   Generally speaking, infiltration refers to water that moves into the soil, although it may never reach the saturated zone because of evaporation or transpiration.

Interbed   A comparatively thin bed of rock or sediment occurring between layers of a different material.

Intrusive rock   A type of igneous rock formed as molten material cools in the subsurface. Because intrusive rock cools relatively slowly it tends to have larger mineral crystals than extrusive (volcanic) rock. Synonymous with irruptive rock.

Lacustrine   Deposits or geologic features formed in or related to lakes.

Limestone   A sedimentary rock composed of the mineral calcite (calcium carbonate) and varying degrees of fine sediment. Limestones may also contain the mineral dolomite in lesser amounts.

Lithology   The physical and mineralogical characteristics of a rock. Common lithologic names may denote a specific type of rock, for example, sandstone, basalt, or granite, or may denote the general mode of rock formation, for example, sedimentary, volcanic, or intrusive.

Loess   A fine-grained, wind-deposited sediment (windblown dust) rich in clay.

Losing stream   The interaction between a stream and an aquifer can usually be described by whether the stream gains water from or loses water to the aquifer. A gaining stream receives or gains flow from groundwater. A losing stream is one that loses or contributes flow to groundwater.

Mantle plume   An active upwelling of magma from near the core-mantle boundary that significantly affects the Earth's surface (commonly as a hot spot).

Massive   A thickness of stratified sedimentary or volcanic rock without visible internal structure such as beds, joints, or fractures.

Metamorphic rock   Rock formed by the alteration of existing rock by heat and pressure.

Metasiltstone   A lightly metamorphosed siltstone.

Micrite   Sedimentary rock in which the carbonate mineral crystals are less than 4 micrometers in diameter; a very fine-grained limestone or dolostone.

Mineral   A solid crystalline substance formed by natural processes. These processes are usually inorganic but may be organic. Minerals are the building blocks of rocks and sediment.

Mineralization   The introduction of minerals into a rock resulting in an economic or potentially economic ore deposit.

Multiple-well aquifer test   A type of aquifer test in which aquifer qualities are inferred by pumping a single well at a known rate and observing drawdown over time in an adjacent non-pumping well or wells. Because multiple wells are used the influence of well construction is minimal, thus multiple-well tests better determine aquifer properties than single-well tests.

Normal fault   A type of dip-slip fault in which the upper (or hanging) wall has slipped down relative to the lower (or footwall).

Olivene   A hard, green, yellow, or brown, prismatic silicate mineral. It is usually formed in igneous rock and is an important constituent of such rocks as basalt and gabbro.

Orogeny   Mountain building; the process by which rocks are folded, thrust-faulted, metamorphosed, and intruded followed by uplift.

Outcrop   An area of exposed bedrock.

Overbank deposits   Sediments deposited on the flood plain of a river outside the main channel. They tend to be fine-grained because slower-moving water on the floodplain will not transport larger sediment.

Pahoehoe   A basaltic lava flow with a smooth, ropy surface.

Paleostream   A stream that existed in the geologic past; a common indicator of such a stream is the presence of channels incised into an older surface.

Partially-penetrating well   A well that is drilled through and (or) screened over a section of an aquifer rather than the full thickness.

Pediment   A gently sloping erosional surface at the base of a mountain front. A pediment is an indicator of a retreating mountain front. A pediment is discontinuously mantled with a thin veneer of sediment as it is transported from the eroding upland.

Perennial stream   A stream that flows year-round from either upstream runoff or the contribution of groundwater.

Public Land Survey System (PLSS)   The survey system used to survey much of the United States, it is based on a rectangular grid system using the mile. Divisions are townships, sections, and the successive quartering of sections. It is sometimes referred to as the township and range system.

Porosity   The ratio of openings (or voids) in a rock—such as between sedimentary grains or within fractures—to the total volume of the rock, usually expressed as a percentage. Such openings may not be connected; thus, a rock with significant porosity may have a low value of hydraulic conductivity. In general, the more uniform (or better sorted) the rock material, the greater will be its porosity. Additionally, fine-grained materials tend to be better sorted than coarser materials.

Porosity, primary   Porosity due to the original pore space created during sediment deposition or rock formation.

Porosity, secondary   Porosity formed after sediment deposition or rock formation due to dissolution, weathering, or fracturing.

Porphyry   An igneous rock containing large, conspicuous mineral crystals within a finer-grained matrix.

Potentiometric surface   The level to which water rises in a tightly cased well; it represents the total head of groundwater.

Quartzite   A very hard but unmetamorphosed sandstone composed mostly of quartz grains with a silica cement.

Radiometric dating   The method by which geologic materials are assigned ages based on the measurement of radioactive parent and daughter isotopes.

Recharge   Water that ultimately enters the saturated zone and thus contributes water to the aquifer.

Rhyolite   A typically light-colored extrusive igneous rock with the same general chemical composition as its intrusive equivalent, granite. Rhyolitic magmas have a very high viscosity and eruptions tend to be highly explosive affecting large areas (such as the Yellowstone caldera).

Rock units   Rock units may be classified and mapped on the basis of many different criteria including lithology, magnetic polarity, age, and depositional environment. The most common method is to classify rock strata of about the same age and similar physical characteristics into formations. Formations may be subdivided into members and beds or aggregated into groups. By convention, the formal name of the unit is a geographic feature near the type exposures (or outcrops) of the rock unit. For example, the Milligen Formation is named for exposures in Milligen Gulch near Triumph, Idaho.

Sand   A rock fragment or detrital particle with a diameter between 0.0025–0.08 inch. Grade names are very fine, fine, medium, coarse, and very coarse.

Sandstone   A medium-grained consolidated sedimentary rock composed primarily of sand-sized sediment with a calcium carbonate, silica, or iron oxide cement. Sandstone may have a fine-grained matrix of silt and clay.

Saturated thickness   The thickness of the aquifer saturated with water. In an unconfined aquifer, the saturated thickness varies with the position of the water table.

Sedimentary rock   Rock formed by the aqueous deposition of sediment or by chemical precipitation.

Sedimentary particle size   Sedimentary particles are generally described according to their grain size. The classification system used in this report is the Modified Wentworth Scale (Ingram, 1989).

Seismic   Relating to earthquakes or other vibrations of the Earth; such vibrations may have a natural or artificial source. Seismic geophysical methods involve the recording and analysis of the different characteristics of seismic waves in order to characterize the subsurface.

Seismometer   An instrument that detects movement in the Earth.

Shale   A fine-grained sedimentary rock, formed by the consolidation of clay and (or) silt.

Siliceous   A rock containing abundant silica; it may be sedimentary, igneous, or metamorphic and the silica may have either an organic or inorganic source.

Silt   A rock fragment or detrital particle with a diameter between 0.00016-0.0025 inch. Grade names are very fine, fine, medium, and coarse. A mixture of silt and clay is mud.

Siltstone   A fine-grained consolidated sedimentary rock with a predominance of silt over clay.

Single-well aquifer test   A type of aquifer test in which aquifer qualities are inferred from the behavior of the water level in a single well. While single-well tests include non-pumping methods such as a slug test, they commonly involve pumping a well at a known rate over a given period of time and observing the drawdown within the well. Single-well tests are affected by wellbore storage and well construction and thus may be a better indicator of well efficiency rather than aquifer properties.

Specific capacity   A measure of well productivity derived by dividing discharge rate by drawdown of water level in the well. Units are length/time.

Stock   An irregularly shaped body of intrusive igneous rock with an exposed surface less than about 40 mi^2; a smaller form of batholith.

Storage coefficient   The volume of water taken into or released from storage (volume) per unit of surface area of the aquifer (area) per unit change in head (length); units simplify to a dimensionless value. Synonymous with storativity and coefficient of storage.

Stream aggradation   The process by which the channel of a stream is raised by sediment deposition within the channel. Stream aggradation typically leads to lateral channel migration and may be caused by such things as climatic changes and consequent changes in the stream's flow regime, tectonism, volcanism, and the formation of downstream lakes.

Stream stage   The height of the stream's water surface, above an arbitrarily established datum plane at which the stage is zero. Synonymous with gage height.

Stream terrace   Steplike benches that parallel a stream and represent different climatic and geologic episodes in the stream's history.

Streamflow-gaging station   The installations used by hydrologists to monitor the flow of water in streams and rivers. Gaging stations typically consist of a shelter that encloses a recorder to monitor the height (or stage) of the stream's water surface. A correlation (known as a rating curve) can be made between stage and stream discharge by periodically measuring the streamflow rate (or discharge) of the stream and comparing it to the stage. An increasing number of recorders in gaging stations broadcast their stage data in real time or near real time by satellite or telephone. These data are used to automatically calculate discharge, and the discharge is then made available over the Internet. This streamflow information is useful not only for resource management and flood warning but also for recreational purposes such as fishing and boating.

Strike   The direction of a geologic surface (such as a fault plane or bedding) as it intersects the horizontal.

Tectonic plate   A thin, rigid, segment of the Earth's crust that moves horizontally; about 15 of these plates make up the surface of the Earth. The boundaries between these plates are areas of seismic, volcanic, or tectonic activity.

Tectonism   Movement of the Earth's crust related to the formation of such large-scale features as ocean basins, continents, mountain ranges, or plateaus.

Thrust fault   A dip-slip fault with an acute dip in which the upper block overrides the lower block, in some cases for great distances.

Total head   In groundwater, the sum of the elevation, pressure, and velocity heads. Total head is represented by the potentiometric surface.

Transmissivity   A measure of the rate at which a porous medium transmits water through a unit width of the aquifer. It is equal to hydraulic conductivity times the saturated thickness of the aquifer.

Tuff   A volcanic rock composed of volcanic ash and larger volcanic ejecta; it may be strongly to weakly consolidated.

Unconfined aquifer   An aquifer in which the water table is exposed to the atmosphere through openings in the overlying materials.

Underflow   Subsurface flow beneath or adjacent to a stream channel.

Vesicle   A void in an igneous rock formed by a bubble of gas.

Volcanic rock   A type of igneous rock formed as molten rock, ash, or volcanic ejecta cools on the surface. Because it cools relatively quickly volcanic rock tends to have much smaller mineral crystals than intrusive rock. Synonymous with extrusive rock.

Volcaniclastic rock   A rock composed solely or mostly of fragments of volcanic rock transported and deposited by wind or water.

Well-performance test   A type of single-well aquifer test to determine the maximum rate that a well can be pumped along with the corresponding drawdown in order to identify the most efficient combination of pump capacity and depth.

Appendix A. Well Information Used in the Estimation of the Altitude of the Pre-Quaternary Bedrock Surface and Top of Quaternary Basalt, Wood River Valley Aquifer System, South-Central Idaho

Data are available for download at http://pubs.usgs/gov/sir/2012/5053/.

Appendix B. Horizontal-to-Vertical Spectral Ratio Measurements Used to Estimate Altitude of the Pre-Quaternary Bedrock Surface and Top of the Quaternary Basalt, Wood River Valley Aquifer System, South-Central Idaho

Data are available for download at http://pubs.usgs/gov/sir/2012/5053/.

Appendix C. Well Information For Wells Not Completed to the Pre-Quaternary Bedrock Surface or Top of Quaternary Basalt, Wood River Valley Aquifer System, South-Central Idaho

Data are available for download at http://pubs.usgs/gov/sir/2012/5053/.

Appendix D. Well Information Used in the Estimation of the Top and Corresponding Thickness of the Uppermost Unit of Fine-Grained Sediment Within the Wood River Valley Aquifer System, Southern Wood River Valley, South-Central Idaho

Data are available for download at http://pubs.usgs/gov/sir/2012/5053/.

Appendix E. Estimates of Transmissivity and Hydraulic Conductivity for 81 Wells in the Wood River Valley, South-Central Idaho

Data are available for download at http://pubs.usgs/gov/sir/2012/5053/.